SURVEY OF ORGANIC CHEMISTRY STUDENT PRIMER 2018

BY TANIA HOUJEIRY AND THOMAS D. HICKMAN

Marketed by ProtonGuru.com

Find additional online resources and guides at protonguru.com, a new site under construction as of summer 2018!

There is a lot of online organic chemistry video content on the Proton Guru YouTube Channel! Just go to YouTube and search "Proton Guru Channel" to easily find our content.

Questions for the editorial staff can be addressed to IQ@protonguru.com

Cover photo courtesy of Ashlyn D. Smith

Printed in the United States of America

10 9 8 7 6 5 4 3 2 1

ISBN 978-1721899883

Survey of Organic Chemistry Student Primer

2018

**By Tania Houjeiry, Ph.D.
and Thomas D. Hickman, Ph.D.**

Table of Contents

Purpose of the Primer

This brief, plain language Primer is meant to provide a survey of the most basic principles and concepts of organic chemistry presented as simply as possible. It is meant to be an excellent *primer* to read *before* lecture so that once you get to lecture you will already have some knowledge of what will be discussed. It is also meant to be an excellent pre-exam review. When paired with practice problems and handouts or video tutorials, this student primer takes the place of an expensive, bulky textbook for a one-semester organic chemistry survey.

Online Content

There are many resources to support organic chemistry study online. One good resource for these resources is protonguru.com (launched in summer 2018). This is an ever-developing arsenal of study resources that will grow over time, so check back throughout fall 2018 and beyond!

Lesson 1. Atoms and Bonding Review

Lesson 1.1. Valence and Types of Bonds

Organic chemistry is the study of compounds containing carbon, and often containing hydrogen, nitrogen, and oxygen. From general chemistry, we know that carbon is atomic symbol "C" on the periodic table. We can deduce the electronic configuration of C from its position on the periodic table. Let's review a bit about electronic configurations here. Carbon's electron configuration is $1s^2 2s^2 2p^2$ and orbital diagram are shown below.

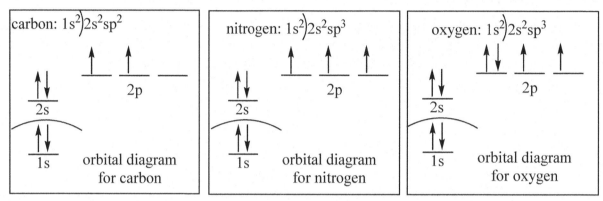

The $n = 1$ shell is filled so it's inert with respect to reactivity. We can focus on $n = 2$ shell for valence electrons and reactivity.

Carbon has 2 core electrons and 4 valence electrons; nitrogen has 5 valence electrons; oxygen has 6 valence electrons. We should recall from general chemistry that atoms have different attraction for electrons. The pull an atom has for electrons in a bond is referred to as the atom's **electronegativity** (EN). Electronegativity increases on the periodic table across a row and up a column, basically as we move towards the most electronegative element of fluorine.

An arbitrary scale of EN values (referred to as the Pauling electronegativity) is often used in general chemistry to determine if a bond is either ionic or covalent. Fluorine is given the value of 4.0 and other elements are assigned proportional values less than 4.0. In bonding situations, a difference in electronegativity between the two bonded atoms greater than 2.0 suggests that we have an ionic bond, a difference between 0.5 and 1.9 suggests we have a polar covalent bond, and a difference of 0.5 or less says we have a nonpolar covalent bond. Some EN values and the trends are shown below. In terms of electronegativity, one

Some electronegativity values

EN increases →

H 2.1						
Li 1.0	Be 1.5		B 2.0	C 2.5	N 3.0	O 3.5	F 4.0
Na 0.9	Mg 1.2		Al 1.5	Si 1.8	P 2.1	S 2.5	Cl 3.0
K 0.8							Br 2.8
							I 2.5

EN increases ↑

must remember that hydrogen's electronegativity falls in between boron and carbon. This explains why **C—H bonds are nonpolar covalent bonds**.

It should have been stressed in general chemistry that bonding between atoms is a continuum, with pure covalent bonds on one end and pure ionic bonds on the other end of the scale. Most general chemistry books will describe bonds with terms and calculations of "percent ionic character" and "percent covalent character." We don't concern ourselves with such specific distinctions in organic chemistry but we do recognize the difference between ionic and covalent bonds. Ionic bonding often occurs between a metal and a nonmetal. Covalent bonding most often occurs between non-metals. We often come away from general chemistry with these notions:

- Ionic bonds involve the complete transfer of electrons between atoms.
- Covalent bonds involve the sharing of electrons between atoms.

Sometimes a partial periodic table like the one shown below is provided by your organic chemistry professor or on standardized exams so that you can make decisions using it. The table below only shows the atomic number and the atomic mass. Note that the actual EN values are not provided, so we rely on our knowledge of the trends when we need to use EN in a problem.

1A								8A
1 **H** 1.01	2A		3A	4A	5A	6A	7A	2 **He** 4.00
3 **Li** 6.94	4 **Be** 9.01		5 **B** 10.81	6 **C** 12.01	7 **N** 14.01	8 **O** 16.00	9 **F** 19.00	10 **Ne** 20.18
11 **Na** 22.99	12 **Mg** 24.30		13 **Al** 26.98	14 **Si** 28.09	15 **P** 30.97	16 **S** 32.07	17 **Cl** 35.45	18 **Ar** 39.95
19 **K** 39.10							35 **Br** 79.90	
							53 **I** 126.90	

Lesson 1.2. Writing Lewis Dot Structures (LDS)

In general chemistry, you certainly learned how to draw Lewis dot structures and how to interpret them. After a fair amount of practice, you could draw a structure with little thought just by following some familiar steps. Here is a truncated summary of these steps. Review your general chemistry notes for a more thorough treatment.

- Add up all electrons (valence electrons from all atoms and consider charges)
- Decide on molecular sequence and use a pair of electrons to draw a bond between atoms.
- Subtract 2 from your total for each bond you make in your structure. The least EN atom is usually the central atom.
- Distribute remaining electrons, in pairs, to more electronegative peripheral atoms so each has 8 electrons surrounding it (H only gets 2 electrons). Subtract each pair from your running total.
- Place any remaining electrons on the central atom.
- Satisfy the octet rule; but know that some exceptions exist.
- You can move a lone pair of electrons from a neighboring atom that doesn't yet have 8 electrons, resulting in a multiple bond.
- Boron only needs 6 electrons but *can* hold 8.
- Hydrogen will have 2 electrons.
- Atoms beyond the second row can have more than 8 valence electrons.
- Calculate formal charges in your evaluation of the structure.

We can be more efficient with our organic Lewis dot structures when we consider the Lewis symbols for the most common atoms and how we can relate an atom's Lewis symbol to the number of bonds the atom has in a usual neutral form. If you place valence electrons around the atom one at a time and place them on one of the four sides (top, bottom, left, right) at a time, you can anticipate what the preferred bonding scenario is for each atom.

These are the most common atoms bonded to carbon and these bonding scenarios keep the formal charge of the atom at zero:

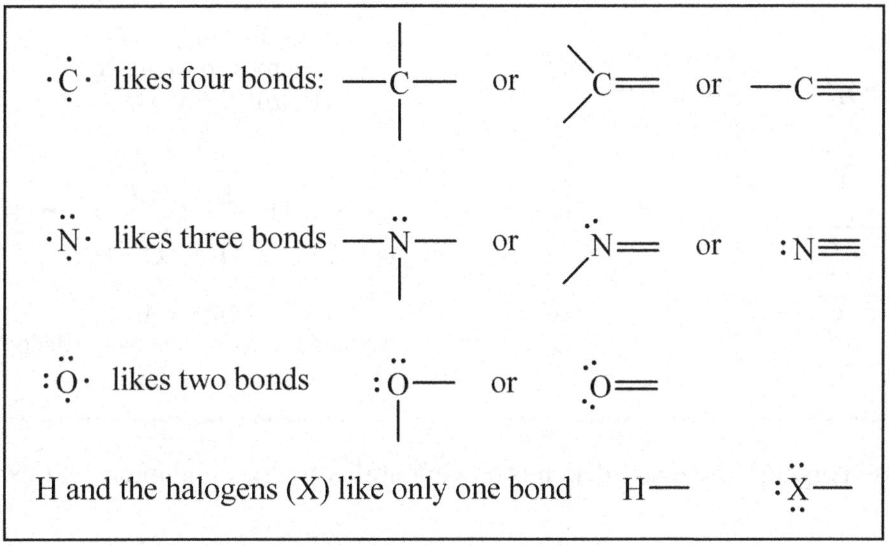

Here are two examples of Lewis structures being drawn with the formal procedure and our knowledge of how these common atoms like to bond.

Example 1.1

Draw the Lewis structure for a molecule having the formula a) CH_2Cl_2 and b) CH_3COOH.

Solution 1.1

a)

CH_2Cl_2

C	4 v.e.'s
2 H	2
2 Cl	14

$\underline{}$
20
-8
$\overline{12}$
-12
$\overline{0}$

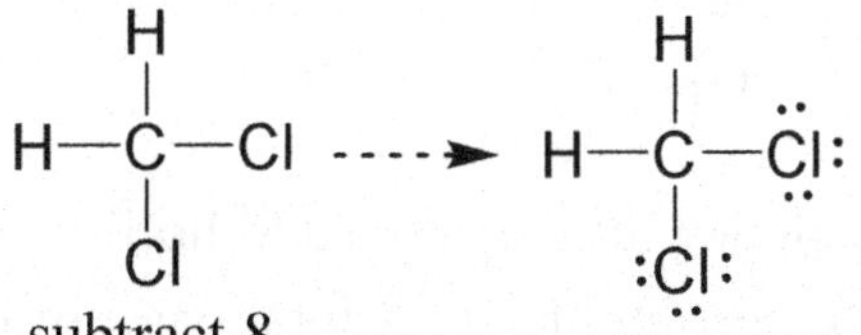

subtract 8
for the 4 bonds
we made

subtract 12
for the six lone pairs

H-atoms and halogens prefer one bond
and all atoms have their quota of electrons.
We are finished.

b)

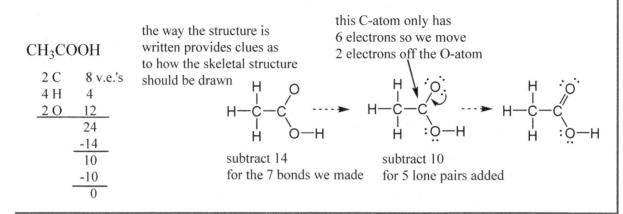

CH₃COOH

2 C   8 v.e.'s
4 H   4
2 O   12
        24
        -14
        10
        -10
        0

the way the structure is written provides clues as to how the skeletal structure should be drawn

this C-atom only has 6 electrons so we move 2 electrons off the O-atom

subtract 14 for the 7 bonds we made

subtract 10 for 5 lone pairs added

Your success in organic chemistry depends on your ability to draw and interpret Lewis dot structures rapidly.

Lesson 1.3. Representing Structures

Drawing Lewis structures has evolved over time and understanding condensed structural formulae and line-bond structures will be difficult in the beginning, but you will soon see how much cleaner and more efficient these structures are. Let's look at the molecule known as Lipitor, which is used to reduce cholesterol. If all the carbon atoms and hydrogen atoms were included, it would be a mess. Line-bond formulae allow our eyes to focus in on the important aspects of the molecule.

Lipitor

Lewis dot structures originally included all electrons and no lines were drawn to represent bonds. They evolved into using the lines but still included all lone pairs of electrons. Kekulé structures are basically Lewis structures without the lone pairs drawn; we're just supposed to know they are there. We are supposed to know this based on our earlier discussion of common bonding scenarios for H, C, N, O, and halogens based on them preferring a formal charge of zero.

$$H\!:\!\overset{\overset{\displaystyle H}{..}}{\underset{\underset{\displaystyle H}{..}}{C}}\!:\!\overset{\overset{\displaystyle H}{..}}{\underset{\underset{\displaystyle H}{..}}{C}}\!:\!\overset{..}{\underset{..}{O}}\!:\!H$$

old school
Lewis dot structure

$$H-\overset{\overset{\displaystyle H}{|}}{\underset{\underset{\displaystyle H}{|}}{C}}-\overset{\overset{\displaystyle H}{|}}{\underset{\underset{\displaystyle H}{|}}{C}}-\overset{..}{\underset{..}{O}}-H$$

newer
Lewis dot structure

$$H-\overset{\overset{\displaystyle H}{|}}{\underset{\underset{\displaystyle H}{|}}{C}}-\overset{\overset{\displaystyle H}{|}}{\underset{\underset{\displaystyle H}{|}}{C}}-O-H$$

Kekule'
structure

CH_3CH_2O

condensed
structural
formula

Condensed structural formulae allow us to conveniently type some, but not all, stru[...]
fashion. Confusion sometimes arises when halogens or parentheses are incorporated.

$CH_3CH(CH_3)CH_2OH$ implies $CH_3-\overset{\overset{\displaystyle CH_3}{|}}{CH}-CH_2-OH$ 　　$(CH_3CH_2)_2NH$ implies 　CH_3CH_2 / CH_3CH_2

In the line-bond notation, a line represents a bond, just as it does in the Lewis structure nota[...]
However, in the line-bond notation, each bend in a line, end of a line, or change in the number of li[...]
at a junction (e.g., single bond to triple bond) represents a carbon atom. In this notation, one assume[...]
the correct number of H atoms on each carbon for the given formal charge. A neutral carbon atom has
four bonds, so any bonds not explicitly shown in the line-bond notation are assumed to be to H; one
does not need to draw out the H atoms on carbon atoms whose symbols are not drawn out as a C. All
other non-C atoms, as well as any H atoms attached to non-carbon atoms must be shown. Also, if you
draw out a 'C' letter for a carbon atom by choice (even though you do not have to), you would need to
show all the H atoms on that C, for example:

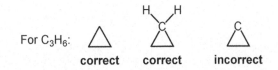

For C_3H_6: 　　correct　　correct　　incorrect

Here are a few examples of compounds shown in both Lewis structure and line-bind notation. One
can readily see how the line bond notation can save a lot of time in drawing a structure as well as provide
a simplified, compact picture of the molecule:

represents the same molecule as:

represents the same molecule as:

represents the same molecule as:

ctures in a linear

you to see atoms other than carbon without getting lost in all the hydrogen

Some atoms can take on a full positive or negative charge, as you have no doubt seen in ions in your prior courses in chemistry. It is important to know how to assign formal charges in organic chemistry. The formula for formal charge of an atom in a structure can be given as follows:

Formal charge =
[# of valence e^- in neutral atom] – [# nonbonding e^- on the atom] – [number of bonds to that atom]

Consider carbon in several different structures:

(valence - nonbonding electrons - bonds)	(valence - nonbonding electrons - bonds)
(4 – 0 – 4) =	(4 – 0 – 3) =
formal charge of 0 (zero) on C	formal charge of +1 on C
(valence - nonbonding electrons - bonds)	(valence - nonbonding electrons - bonds)
(4 – 2 – 2) =	(4 – 2 – 3) =
formal charge of 0 (zero) on C	formal charge of –1 on C

In the structures above, all the bonds and non-bonding electrons are shown. A neutral carbon atom has four valence electrons. One simply plugs the values into the equation and the formal charge is readily determined. It is vitally important to determine which atoms in a molecule have formal charges, so that we can begin to predict the properties and reactions of molecules as a result of Coulombic attraction or repulsion. Learning the ability to make such predictions forms the majority of an introductory organic chemistry course.

If you are given a structure in which the carbon atom only has three bonds, you should be very suspicious and know that you are dealing with a radical, a carbocation (+), or a carbanion (–).

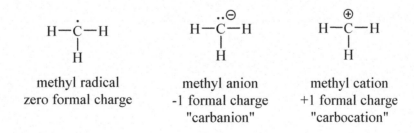

methyl radical	methyl anion	methyl cation
zero formal charge	-1 formal charge	+1 formal charge
	"carbanion"	"carbocation"

If you are given a structure that has a nitrogen atom with more of fewer than its common three bonds or an oxygen with more or fewer than its common two bonds, you should also look for charge or radicals:

:Ö—H

hydroxyl radical
zero formal charge

⊖ :Ö—H

hydroxide ion
-1 formal charge

⊕Ö—H

an oxonium ion
+1 formal charge

N̈=O

nitric oxide is a
nitrogen free radical

H—N̈:⊖
  |
  H

the amide ion is where
the N-atom carries a
-1 formal charge

      H
      |
H—N⊕—H
      |
      H

ammonium ion
N-atom has 4 bonds
and a +1 formal charge

Example 2.1

Provide all non-zero formal charges for atoms in these structures:

A) H₃C—Ö: B) HC≡C: C) :Br: D)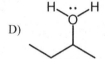

Solution 2.1

A) H₃C—Ö:⁻¹
For O:
6 valence – 6 lone pair electrons – 1 bond
= F.C. of –1

B) HC≡C:⁻¹
For C:
4 valence – 2 lone pair electrons – 3 bonds
= F.C. of –1

C) :Br:⁺¹
For Br:
7 valence – 4 lone pair electrons – 2 bonds
= F.C. of +1

D) H..⁺¹H O
For O:
6 valence – 2 lone pair electrons – 3 bonds
= F.C. of +1

Bond-line formulae make inspection of a molecule easier, but it's important not to forget about the hydrogen atoms that are not drawn. Valence shell electron pair repulsion (VSEPR) theory was developed to explain the observed shapes of molecules.

Each element in the second row of the periodic table has one $2s$ orbital and three $2p$ orbitals in its valence shell. The valence shell electrons are the only ones we show in Lewis structures because they are the only ones involved in bonding and chemical reactions. To accomplish the most effective bonding and minimize repulsion between bonding pairs, the s- and p-orbitals present on *individual atoms* mix to form hybrid orbitals in *molecules*. The s orbital is always mixed into hybrid orbitals. The number of p-orbitals that mix with the s-orbital on a given atom depends on the sum of lone pairs and atoms attached to that atom as follows:

If (lone pairs + atoms attached) = 2   **then**   hybridization = sp

If (lone pairs + atoms attached) = 3   **then**   hybridization = sp^2

If (lone pairs + atoms attached) = 4   **then**   hybridization = sp^3

Note that the total number of orbitals (s + p) that mix to form the hybrid orbitals is equal to the sum (lone pairs + atoms attached).

An sp-hybridized atom has two sp-orbitals on it. To minimize repulsion between the electrons in these orbitals, the orbitals are oriented 180° apart, in a linear geometry:

The two p-orbitals that were not incorporated into the hybrid are still on the C:

The hybrid orbitals may hold lone pairs or form bonds. The type of bond formed by a hybrid orbital involves electron density directly between the two atoms involved in the bond. These are called σ-bonds (σ is the lowercase Greek letter "sigma"). There can be no more than one sigma bond between any two given atoms, and each single bond is a σ bond.

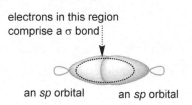

electrons in this region
comprise a σ bond

an *sp* orbital   an *sp* orbital

Each *p*-orbitals that does not mix into the hybrid orbital may interact with a *p*-orbital on an adjacent atom, forming a bond with electron density above and below the internuclear axis. Such a bond is called a π-bond (π is the lowercase Greek letter "pi"). Each bond beyond the first bond between two atoms (which is a σ bond) is a π bond. So, a double bond is made of one σ- and one π-bond.

One π bond:
Two total e⁻ spread over the whole area
above/below internuclear axis (dashed line)

The *sp*-hybridized atom in particular has two *p*-orbitals on it, so it can make two π-bonds:

Molecules with sp hybridized atoms (bold):

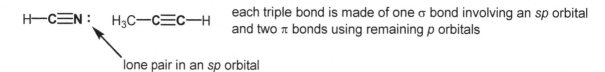

each triple bond is made of one σ bond involving an *sp* orbital and two π bonds using remaining *p* orbitals

lone pair in an *sp* orbital

It is important to know whether electrons are in a lone pair, σ-bond or π-bond because, for a given atom, lone pair electrons are held less strongly (attraction to only one nucleus) than bonding electrons (attracted to two nuclei). Furthermore, the π-bond electrons are held less tightly than the σ-bonding electrons. This knowledge will help us predict which electrons might be the easiest to pull away from a molecule. Thus, for a given atom, lone pair electrons would be easiest to pull away, followed by π-bonding electrons, and σ-bonding electrons would be most difficult to remove. We would therefore predict that, if H⁺ (a species called an "electrophile" that often pulls electrons away from other molecules) interacts with CH_3OH, H⁺ would most likely interact with the lone pair electrons on O. If H⁺ interacted with a molecule having a C=C bond, like $H_2C=CH_2$, the H⁺ would pull the π-bond electrons from the alkene. If H⁺ interacted with $H_3C–CH_3$, however, the σ-bond electrons might be too tightly held to be pulled away by the H⁺. These simple predictions all turn out to be true!

An *sp²*-hybridized carbon atom has three *sp²*-orbitals on it. To minimize repulsion between electrons in these orbitals, the orbitals are 120° apart, in a trigonal planar geometry:

There will be one unhybridized *p*-orbital, so an *sp²*-hybridized atom can make three σ-bonds and one π-bond. Here are some examples of molecules which incorporate *sp²*-hybridized atoms:

Molecules with *sp²* hybridized atoms (bold):

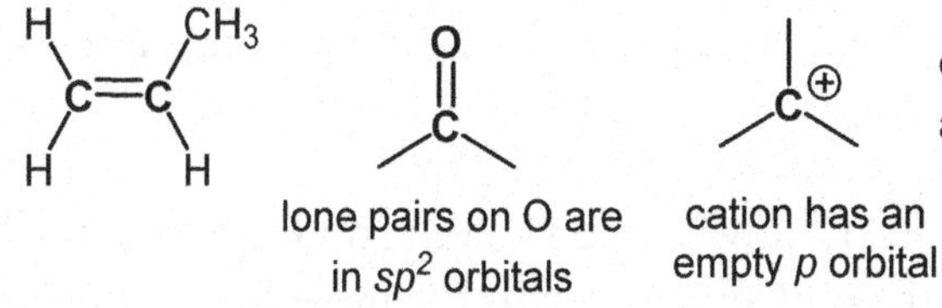

H₂C=CH-CH₃ (with labels H, CH₃, C=C, H, H)

each double bond is made of one σ bond involving an *sp²* orbital and one π bond using the unhybridized *p* orbital

lone pairs on O are in *sp²* orbitals

cation has an empty *p* orbital

An *sp³*-hybridized carbon atom has four *sp³*-orbitals on it. To minimize repulsion between the electrons in each orbital, the orbitals are 109.5° apart, in a tetrahedral geometry:

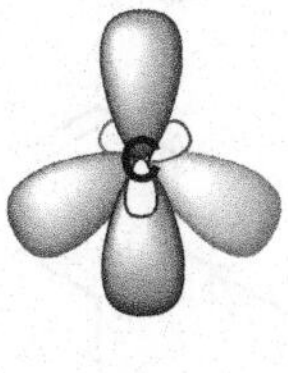

There are no unhybridized *p*-orbitals on an *sp³*-hybridized atom, so it cannot form any π-bonds. The example below illustrates how atom hybridization states are identified in more complex molecules.

Example 3.1

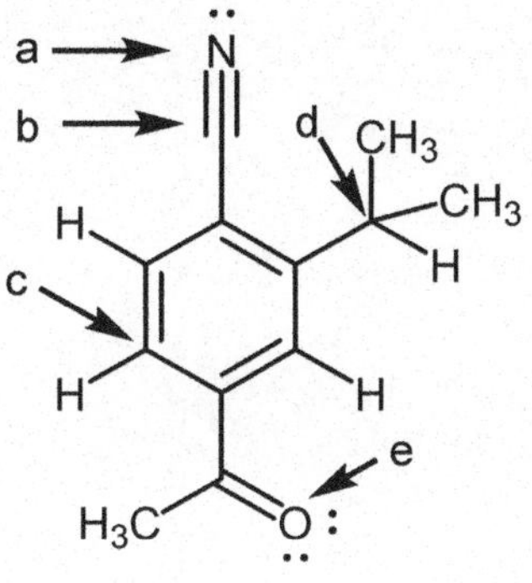

Solution 3.1:

Nitrogen atom (a) has a lone pair and is bound to one other atom; (lone pairs + atoms attached) = 2, so hybridization = *sp*.

Carbon atom (b) is bound to two other atoms and has no lone pairs; (lone pairs + atoms attached) = 2, so hybridization = *sp*.

Carbon atom (c) is bound to three other atoms and has no lone pairs; (lone pairs + atoms attached) = 3, so hybridization = *sp²*

19

Carbon atom (d) is bound to four other atoms and has no lone pairs; (lone pairs + atoms attached) = 4, so hybridization = sp^3

Oxygen atom (e) is bound to one other atom and has two lone pairs; (lone pairs + atoms attached) = 3, so hybridization = sp^2

Note that lone pairs are not always shown on atoms. Likewise, the H atoms on C are not always shown in the line-bond notation. The structure of the molecule given in Example I.6.1 could also have been provided as:

Lesson 4. Electronegativity and Functional Groups

Lesson 4.1. Electronegativity and Polar Bonds

In your prior chemistry courses, you learned about electronegativity (EN) and its influence on polarity, dipole moments, etc. We will briefly review how some of these concepts apply specifically to organic chemistry. As we saw in Lesson 1, EN is the pull an atom exerts on electrons in a bond. A higher value of EN indicates a greater pull for electrons, with EN of an element increasing towards the top and towards the right side of the periodic table so that F is the most electronegative element.

In a bond between two atoms of equal electronegativity, the electrons in the bond will be held exactly in between the two nuclei in a symmetric distribution of charge, so there are no regions of unusually high negative or positive charge, and thus the bond is *nonpolar*. If the electronegativity values for two atoms are quite similar, then those bonds are not significantly polar. *An important example of nonpolar bonds in organic chemistry are C–H bonds.*

Examples of Nonpolar Bonds

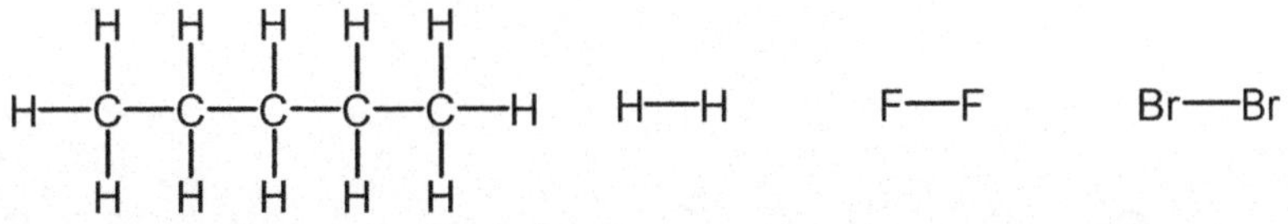

ALL the bonds in this molecule!

If one of the atoms in a bond has a significantly higher electronegativity, it pulls the electrons more closely towards it, so that there is more negative charge on its end of the bond, and less on the other end from which the electrons were pulled, creating a *polar bond*:

Arrows Point to Polar Bonds in these Examples

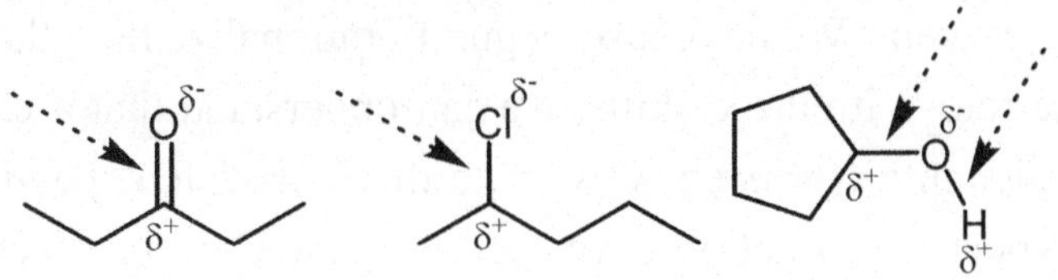

In the picture above, you may notice two new symbols. The "δ–" symbol indicates the *partial negative charge* present at the negative pole of a *polar bond*, while the "δ+" symbol indicates the *partial positive charge* present at the positive pole of the polar bond. The use of these symbols to indicate what sites on a molecule have an excess of charge will become very important to your ability to solve problems as we move through organic chemistry. The greater the difference in electronegativity between the two bonded atoms, the more polar the bond will be, and thus there will be greater amounts of positive and negative partial charges on each side of the polar bond.

The polarity of a bond may also be represented by a *dipole moment* arrow. These arrows are vector arrows that have a "plus sign" at one end and the arrow head points in the direction in which the electrons are pulled:

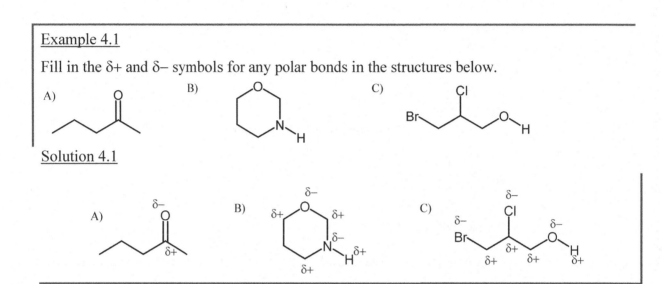

Using Dipole Arrows to Indicate Polar Bonds

Example 4.1

Fill in the δ+ and δ– symbols for any polar bonds in the structures below.

Solution 4.1

Lesson 4.2. Functional Groups

We now have the ability to identify sites in molecules between which there may be partial charges based on which bonds are present. We have also begun to rationalize how these attractive forces could lead to a chemical interaction. With these skills, we can understand that a C=O bond might generally interact with another species in a consistent way, regardless of which specific molecule contains that C=O bond. You would expect the partial positive charge on the C of the C=O unit to attract anions, for example. You might also expect the partial negative on the O in the O–H bond to attract cations, regardless of the molecule on which the OH group resides. We can make generalizations about how a particular **group** of atoms might **function**. Organic chemists have identified specific groups of atoms that do react in predictable manners within a wide range of molecules. These predictably-reacting groups are called **functional groups**. The structures and names of the most important functional groups are shown on the following page. The "R" in these generic structures may be any chain or ring composed of C and H atoms, attached to where the "R" group is placed.

Hydrocarbons

R–C–C–R (with R substituents)
Alkane

C=C (with R substituents)
Alkene

R–C≡C–R
Alkyne

(benzene ring with R substituents)
Aromatics (arenes)

Heteroatom-Containing Compounds

R–N (with methyl groups)
Amine

R–OH
Alcohol

R–O–R'
Ether

R–X
Alkyl halide
(X = F, Cl, Br, I)

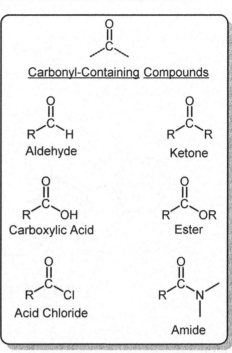

Carbonyl-Containing Compounds

(R–C(=O)–H) **Aldehyde**

(R–C(=O)–R) **Ketone**

(R–C(=O)–OH) **Carboxylic Acid**

(R–C(=O)–OR) **Ester**

(R–C(=O)–Cl) **Acid Chloride**

(R–C(=O)–N) **Amide**

C=O is refered to as a carbonyl bond. The functional groups above contain carbonyl subunits, but "carbonyl" is not itself a functional group.

It is important to be able to look at a complex molecule and rapidly identify all such functional groups that are present. Below is an example of an anti-cancer drug with its functional groups labeled:

Taxol (a complex cancer drug) without (left) and with (right) functional groups labeled

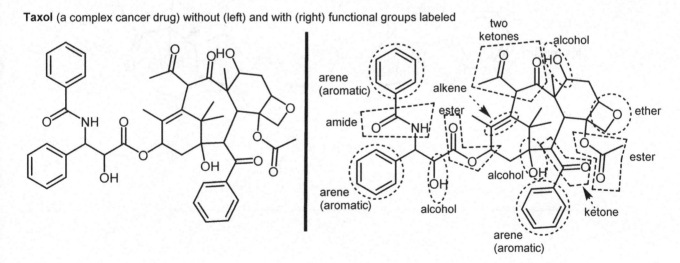

Example 4.2

Identify all non-alkane functional groups in this molecule by circling them and writing in the group's name beside each.

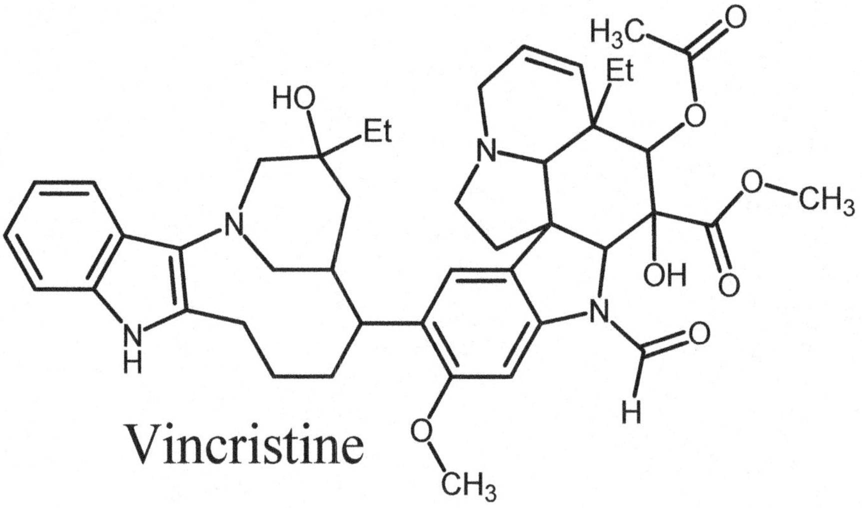

Vincristine

Solution 4.2

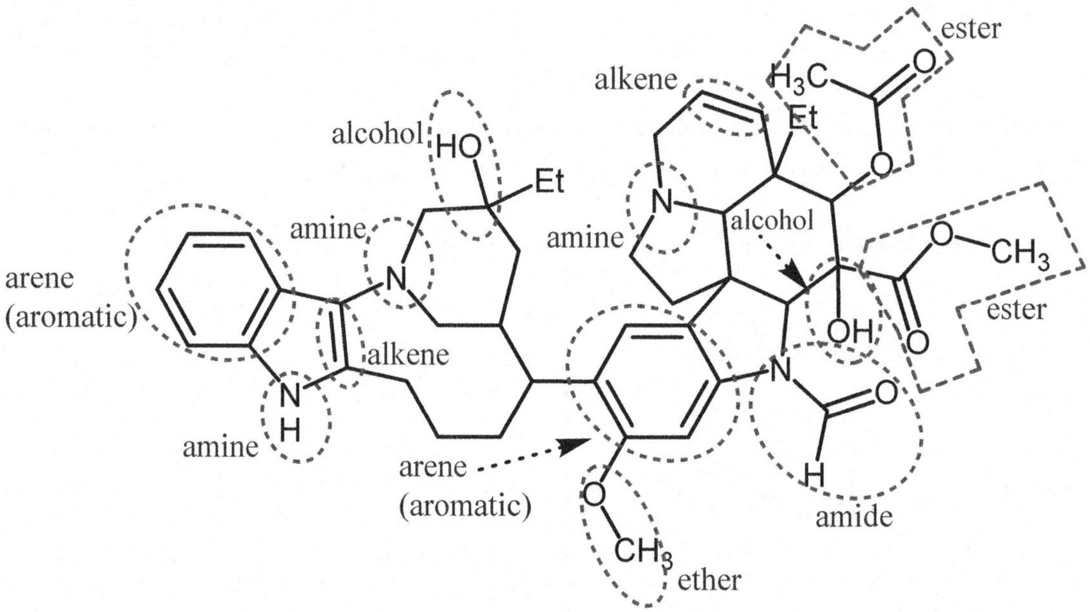

Lesson 5.1. Introduction to the Curved Arrow Formalism in Resonance

In general chemistry, you were probably asked to compare and evaluate two or more resonance structures. In organic chemistry, you will be asked to generate your own resonance structures and this is accomplished by simply following some basic rules and not violating the octet rule. In the beginning it may seem difficult, but it gets easier, even intuitive, with practice.

The *curved arrow formalism* was developed to keep track of valence electrons during chemical reactions where bonds are broken and formed. It's also useful for explaining stabilities of intermediates and why reactions prefer one pathway over another.

We use arrows to show electron movement. When a curved arrow has two barbs it represents the movement of two electrons. If a curved arrow has only one barb then we are moving only one electron. Although there are very complex ways to apply the curved arrow formalism, for now we are only concerning ourselves with how to properly transition from one resonance structure to another.

two barbs, two electrons are moved

one barb, one electron is moved

When we draw the Lewis dot structure for the carbonate ion according to our rules, we can potentially draw three different, equally correct structures:

The carbonate ion drawn three different ways, and all atoms have 8 electrons.

Before discussing how we transition from one form to the other we need to lay down some rules for drawing resonance structures.

1. Only π electrons or lone pair electrons can be shifted, and then only to adjacent atoms or bond positions. Atoms and σ electrons cannot be moved. Here are the three allowed movements:

a. from π bond to an adjacent bond

b. from π bond to an adjacent atom

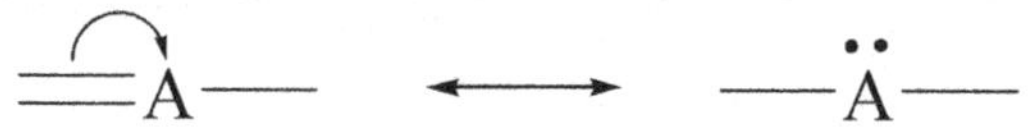

c. from an atom to an adjacent bond

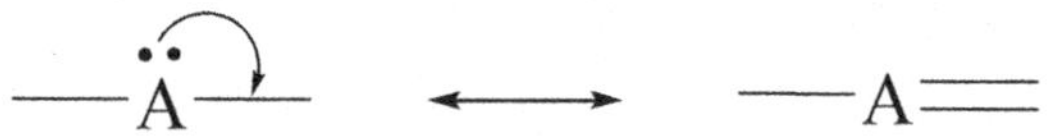

2. Resonance structures in which an atom carries more than its quota of electrons (8 electrons for C, N, O, and F) are <u>not</u> allowed (not real contributors).

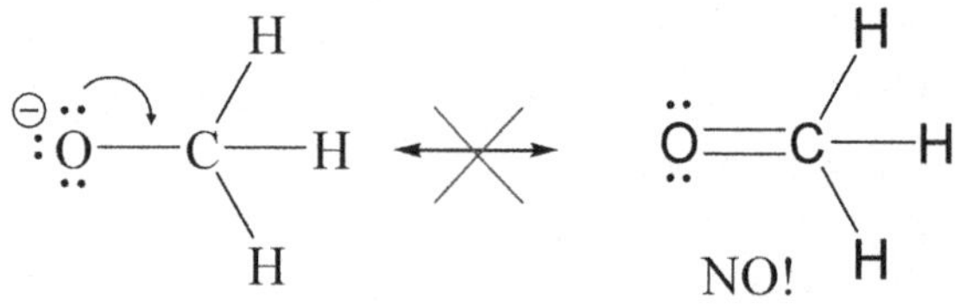

The carbon atom has 10 electrons! This violates the octet rule.

3. The more important (contributing more o the true distribution of charge in the structure) resonance structures show each atom with a complete octet and with as little charge separation as possible. Sometimes resonance contributors are referred to as major or minor contributors to distinguish which Lewis structure is the best approximation of the real structure.

Example 5.1

Rank these resonance contributors in order of importance.

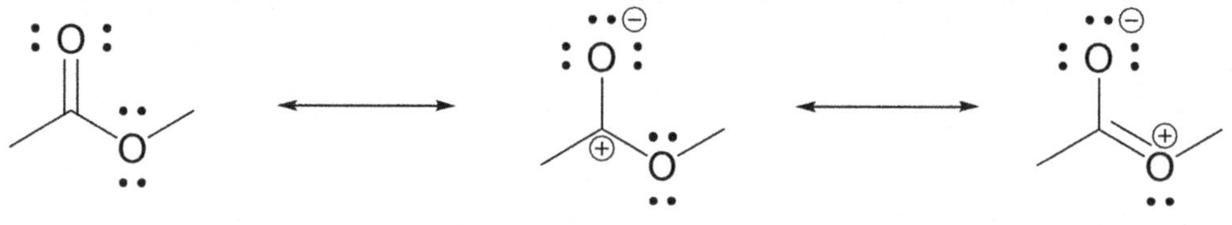

I   II   III

Solution 5.1

Line-bond formulae can make this more trying. Resonance structures **II** and **III** have some formal charge separation and we're supposed to keep this to a minimum. resonance structure **I** has no formal charges and all atoms have their quota of electrons (2 for H, 8 for C and O). It appears that resonance structure **I** is a better resonance structure than resonance structure's **II** and **III**. So how do we evaluate resonance structure's **II** and **III**. We learned in general chemistry that atoms bearing formal charges in resonance structure should reflect the atom's electronegativity. On this basis we might choose resonance structure **II** as the better resonance structure because the O-atom is carrying a negative charge and an O-atom on resonance structure **III** is carrying a negative charge. This would be incorrect because we should note that the positively charged C-atom on resonance structure **II** only has 6 electrons. Having 6 electrons doesn't make resonance structure **II** illegal; it just makes it less important. It's not uncommon for O-atoms to carry a positive charge. So, the correct ranking of importance is **I (most important) > III > II (least important)**.

Lesson 5.2. Common Scenarios in Applying Resonance

Now let's walk through some likely scenarios you will encounter when dealing with resonance structures. THIS is organic chemistry – push/move electrons, check formal charges and evaluate the newly drawn structure. Repeat.

Cations: For a cation, move electrons towards the positive charge. This make sense because a positive charge attracts a negative charge. If you have a positive charge, we would expect electrons to move in that direction. Consider a few examples illustrating correct resonance as well as some common student mistakes:

$$CH_3-CH=CH-\overset{\oplus}{CH_2}$$

Can we generate another resonance structure? What do we do? We're only allowed to move pi electrons or lone pairs and we cannot violate the octet rule. We have 2 pi electrons nearby, so let's try.

$$CH_3-CH\overset{\frown}{=}CH-\overset{\oplus}{CH_2} \longleftrightarrow CH_3-CH-CH=CH_2$$

Okay, we moved the pi electrons towards the (+). The (+) doesn't just disappear. If we move some electrons, we need to check formal charges after.
Look, there is a C-atom with 3 bonds. That is where the (+) went.

$$CH_3-CH\overset{\frown}{=}CH-\overset{\oplus}{CH_2} \longleftrightarrow CH_3-\overset{\oplus}{CH}-CH=CH_2$$

This is correct!

Consider another case:

Can we generate another RS for this cation?

$$CH_3\overset{..}{-\underset{..}{O}}-\overset{\oplus}{CH}-CH_3$$

$$CH_3-\overset{..}{\underset{..}{O}}\overset{\frown}{-}\overset{\oplus}{CH}-CH_3$$

This is NOT a legal movement. We can't move a lone pair to the next atom, only the next bond position.

$$\left[CH_3-\overset{..}{\underset{..}{O}}\overset{\frown}{-}\overset{\oplus}{CH}-CH_3 \longleftrightarrow CH_3-\overset{\oplus}{\underset{..}{O}}=CH-CH_3 \right]$$

Example 5.2

Provide another viable resonance structure for the structure shown:

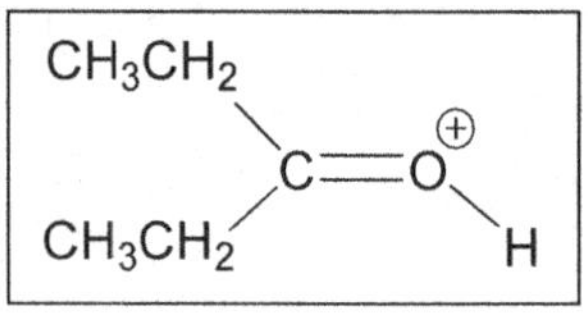

Solution 5.2

Can we generate another RS? It's still just a (+) and we have some moveable electrons nearby. This will utilize one of the other 3 legal electron movements: move pi electrons onto an adjacent atom:

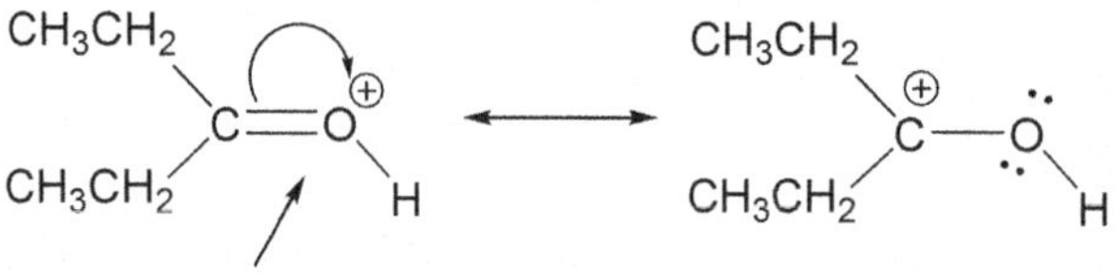

for this O-atom to have
a (+) a lone pair is already
present, just not drawn

If we push some electrons we should check formal charges. The sum of the charges on a species doesn't change. It's not the best RS because the positively-charged C has fewer than an octet of electrons but it is a legal RS.

Anions: For an anion, your first move is to push those electrons (the negative charge) OFF the atom with the (−). This makes sense because we are only allowed to move electrons so we should move those very electrons causing the negative charge if we want to generate a new resonance structure. Consider a few examples illustrating correct resonance as well as some common student mistakes:

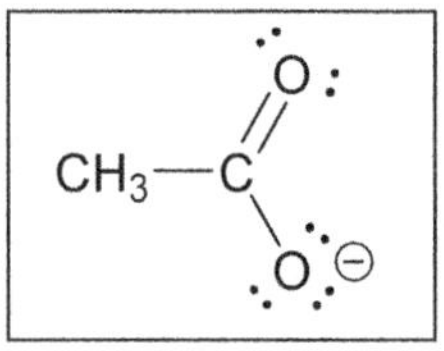

Can we generate another RS for the acetate ion? We were just told that if we have a neg. charge that our first move is to get that (-) off the atom. Let's do it.

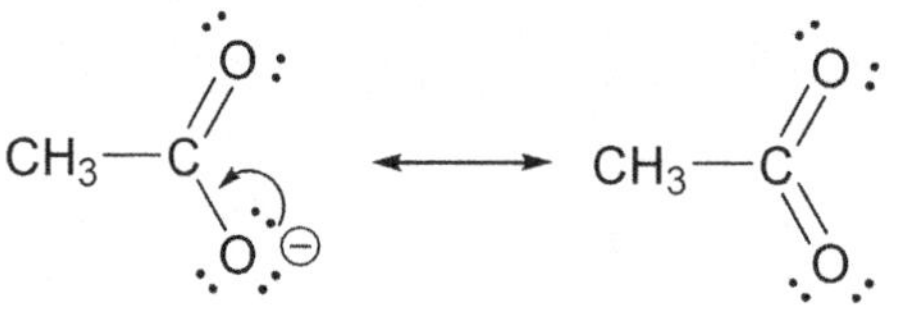

Oh no, that arrow push created a C-atom with 10 electrons and that is definitely not permitted. Are we stuck? Can we do anything?

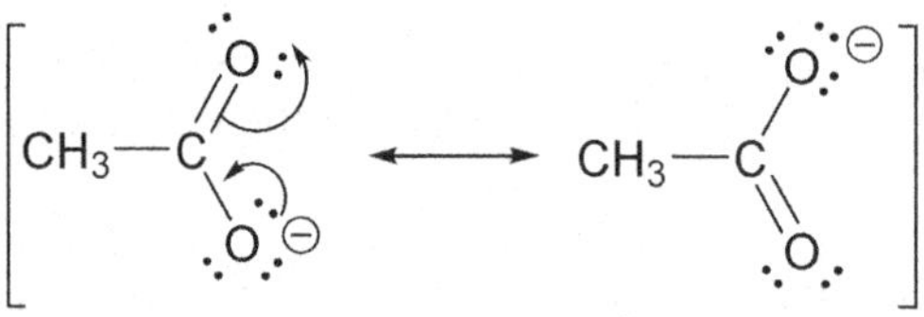

This is correct!

Moving a neg. charge often requires two arrow movements. Lucky for us that C-atom had some pi electrons we could simultaneously move to avoid violating the octet rule.

28

Acetic acid is acidic because its conjugate base, the acetate ion, is said to be resonance stabilized. If you can draw resonance structures then that species is said to be stabilized by resonance.

$CH_3-CH_2-O^{\ominus}$

We have an O-atom with a negative charge and we're not showing any lone pairs. We're just supposed to know that if an oxygen atom has only one bond and a −1 formal charge, then there must be three lone pairs $(6 - ? - 1 = -1)$. The missing 6 implies 3 lone pairs. So we can just move the (−) with arrow and assume it was a lone pair we moved.

$CH_3-CH_2-\overset{..}{\underset{..}{O}}{:}^{\ominus}$ ⟷ $CH_3-CH_2=\overset{..}{\underset{..}{O}}{:}$

Oh, this is NOT allowed because the central carbon now has 10 electrons. What does this mean? Well, it means that another RS cannot be drawn. The negative charge is stuck on that one O-atom. There were no pi electrons to move like in the acetate ion.

$CH_3-CH_2-O^{\ominus}$ For this anion, the negative charge is the sole responsibility of that one O-atom.

What if we are asked to generate more resonance structures from a neutral molecule?

$$\underset{\text{ethyl acetate}}{CH_3-\overset{\overset{\displaystyle :O:}{\|}}{C}-\overset{..}{\underset{..}{O}}-CH_2-CH_3}$$

When asked to generate resonance structures from a neutral molecule the best thing to do is look for a polar bond that has some pi electrons, and that C=O bond fits the bill. We can just push those pi electrons onto the more electronegative O-atom, which makes sense.

$$\left[\quad CH_3-\overset{\overset{\displaystyle :O:}{\|}}{C}-\overset{..}{\underset{..}{O}}-CH_2-CH_3 \quad \longleftrightarrow \quad CH_3-\overset{\overset{\displaystyle ..\ominus}{:O:}}{\underset{\oplus}{C}}-\overset{..}{\underset{..}{O}}{\oplus}-CH_2-CH_3 \quad \longleftrightarrow \quad CH_3-\overset{\overset{\displaystyle ..\ominus}{:O:}}{C}=\overset{..}{O}{\oplus}-CH_2-CH_3 \quad \right]$$

Just one arrow on a polar bond with pi electrons is a good start

Now we know what to do - move electrons towards the (+). Moving the (-) off the O-atom just puts back at the beginning.

We were able to generate two more RS's by just starting with one arrow.

Example 5.3

Generate six more resonance structures for this structure:

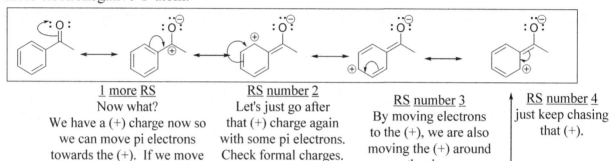

Solution 5.3

How can we get six more RS's from this structure? It doesn't have a (+) charge to move electrons towards; it doesn't have a (-) charge to move off an atom. We do have a polar bond in the C=O bond so our first push could be that of the pi bond electrons up onto the more electronegative O-atom.

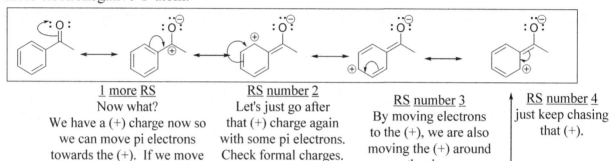

1 more RS
Now what?
We have a (+) charge now so we can move pi electrons towards the (+). If we move the lone pairs on the O-atom, we would simply return to the original structure. Check formal charges.

RS number 2
Let's just go after that (+) charge again with some pi electrons. Check formal charges.

RS number 3
By moving electrons to the (+), we are also moving the (+) around the ring.

RS number 4
just keep chasing that (+).

RS number 6
This looks like the original structure, but it's not. Notice that the pi bonds in the ring have all shifted one position. So it's a new RS. RS 5 looks a lot like RS 1 but it differs in how the pi bonds are situated in the ring.

RS number 5
Now we have to move a lone pair from the O-atom to the (+) charge

Please note that RS 6 and the original are the best resonance structures. This was simply an exercise in drawing resonance structures and how a charge can be spread over numerous atoms in a molecule and not solely belong to one atom.

We prefer the Brønsted-Lowry Theory definition for acids and bases in organic chemistry. Some books will use the broader Lewis Acid-Base definition for acids and bases, particularly when referring to mechanisms, catalysts, and organometallic reactions. Here, we are talking about transferring a proton (H^+) just like you did in general chemistry, and using terms like pH, pKa, conjugate acid-base pair, and applying the Henderson-Hasselbalch equation, as we did in general chemistry. This is also a good opportunity to illustrate how organic mechanisms are drawn. Using our general chemistry knowledge, we already know the outcome of a reaction between an acid and a base, so now it is just a matter of properly applying the curved arrows.

A Brønsted-Lowry **acid** is a proton (H^+) donor; a Brønsted-Lowry **base** is a proton acceptor. An acid may or may not have a positive charge; it just needs to possess an available proton. A base may or may not have a negative charge; it simply needs an unshared pair of electrons.

Two examples:

$$HO^{\ominus} \; + \; H\text{-}Cl \longrightarrow HO\text{-}H \; + \; Cl^{\ominus}$$

$$\text{base} \qquad\quad \text{acid}$$

$$NH_3 \; + \; H\text{-}\overset{\overset{H}{|}}{O}^{\oplus}\text{-}H \longrightarrow \overset{\oplus}{N}H_4 \; + \; H_2O$$

$$\text{base} \qquad\quad \text{acid}$$

We need to review the terms conjugate acid and conjugate base. The **conjugate base** is the ion or molecule formed when an acid loses its H^+. In the above reactions, the Cl^- ion in the first reaction, and the H_2O molecule in the second equation are conjugate bases. The **conjugate acid** is the ion or molecule formed when the base gains a proton. In the above examples, H_2O and NH_4^+ are the conjugate acids in the first and second reactions, respectively.

phenol   methylamine   phenolate ion   methyl-ammonium ion

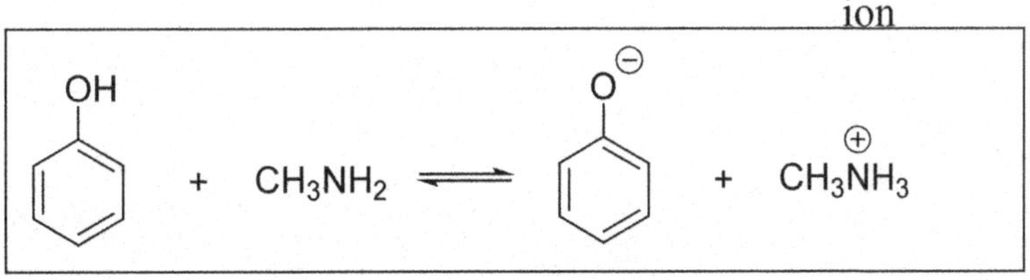

acid   base   conjugate base   conjugate acid

We say phenol and the phenolate ion are a **conjugate acid-base pair**; methylamine and the methylammonium ion are also a conjugate acid-base pair. In other words, the phenolate ion is the conjugate base of phenol; the methylammonium ion is the conjugate acid of methylamine. The reaction is an equilibrium so we can consider the reverse reaction and say that phenol is the conjugate acid of the phenolate ion; methylamine is the conjugate base of the methylammonium ion.

Regardless of the type of acid, scientists use **pK_a** values to quantify acid strength. A lower pK_a value corresponds to a stronger acid. We also know from prior courses that a stronger acid has a weaker conjugate base and that a stronger base has a weaker conjugate acid. This is because chemical reactions are more favorable when more thermodynamically stable species are produced, a theme that we will use to predict reactions throughout organic chemistry. We can conclude that *the more stable the conjugate base anion, the stronger the acid.* Conversely, *the less stable the conjugate base anion*, the greater the favorability for it to be neutralized by an acid, so *the stronger a base it is.* Here is a small pK_a table to illustrate.

	Acid	pK_a	conjugate base	
strongest acid (lowest pK_a)	HCl	-7	Cl^-	weakest base
	H_3O^+	-1.7	H_2O	
	H_2O	15.7	OH^-	
	$HC\equiv CH$	25	$HC\equiv C:^{\ominus}$	
weakest acid (highest pK_a)	NH_3	~ 35	NH_2^-	strongest base

Chemists use pK_a values to predict the relative strengths and stabilities of acids and bases. From the table we can see that OH^- is a stronger (less stable) base than the weaker (more stable) Cl^- ion. The pK_a values can also predict the outcome of acid-base reactions and even determine which side of an equilibrium is favored. What if we're not given pK_a values? Can we somehow predict relative acidity or basicity based only on the structure? It turns out we can, and this is what we are going to consider in the next Lesson.

Lesson 7.1. Structural Factors Influencing Acid Strength

In the previous Lesson, pK_a values were invoked to determine acid and base strength: the lower the pK_a, the stronger the acid and thus the weaker the conjugate base. The focus of this Lesson is to use the chemical structure of a compound to *make relative predictions about acid strength.* We will consider each structural effect individually. Without the benefit of pK_a values, the best way to predict acid strength is to consider the conjugate base anion of the acid and how the base carries the negative charge, providing the original acid was neutral. In general, a more diffuse negative charge is a more stable anion; this applies to two of the five factors we will consider (size of anionic atom and resonance). A negative charge more strongly attracted by positive charges is also more stable; this applies to the other three factors we will consider (electronegativity, inductive effect and hybridization).

Electronegativity

Once a proton is removed, we can consider the atom bearing the negative charge. When the atoms carrying the negative charge are in the same row of the periodic table (and are therefore *about* the same size), the anion having the negative charge on the more electronegative atom will be the more stable anion.

on periodic table:

C	N	O	F

$CH_4 \longrightarrow {}^{\ominus}CH_3$ $NH_3 \longrightarrow {}^{\ominus}NH_2$ $H_2O \longrightarrow {}^{\ominus}OH$ $HF \longrightarrow F^{\ominus}$

pKa = 60   pKa = 36   pKa = 15.7   pKa = 3.1

When faced with the actual pK_a values, it's clear that the more electronegative fluorine is the most capable of stabilizing the negative charge.

Hybridization

Let's compare the pK_a values of some hydrocarbons.

compound	pK_a	hybridization	% s-character
CH_3CH_3	~50	sp^3	25 %
$H_2C{=}CH_2$	~45	sp^2	33 %
$HC{\equiv}CH$	~26	sp	50 %

For a carbon atom, electrons in the spherical $2s$ orbitals are closer to the nucleus than $2p$ electrons. This suggests that a carbon atom with more *s*-character will keep its electrons closer to the nucleus and thus alter the electronegativity of the carbon atom. The relative electronegativities of hybridized carbon atoms is: $sp > sp^2 > sp^3$. This makes the C-H bond more polar and increases acid strength. This idea that hybridization can change electronegativity may be new to you, but you can see that it has a significant impact on the acidity of differently-hybridized C atoms!

Size

When comparing acids where the atom losing the proton (the atom that becomes negative) are in the same column of the periodic table but in different rows, the stronger acid will have the negative charge on the larger atom in its conjugate base. The diffuse negative charge makes for a more stable anion/weaker base.

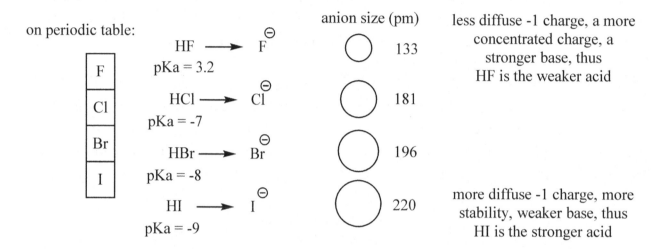

When the negative charge is on a larger atom, the anion is more stable than an analogous species featuring the negative charge on a smaller atom. If the conjugate base anion is more stable, the acid producing that anion is a stronger acid. We conclude that, all else being equal, the stronger acid in a pair of molecules is the one in which the negative charge produced upon deprotonation resides on a larger atom. Consider H_2O vs. H_2S. If H_2O acts as an acid by losing H^+, a negative charge is on O in the conjugate base (^-OH). If H_2S acts as an acid by losing H^+, a negative charge is on a much larger S atom in the conjugate base (^-SH). We would predict that ^-SH is more stable than ^-OH, which would mean that H_2S is a stronger acid than H_2O. The pK_a values confirm our prediction.

To assess anion stability, we do not look at the size of the entire molecule, just the size of the atom bearing the formal negative charge in the conjugate base. Larger atoms in the same molecule are irrelevant to anion stability if they never acquire any negative charge following deprotonation. *Notice that the size is more important than the electronegativity effect* (O is more electronegative than S).

34

Resonance

As just discussed with respect to conjugate base/anion size, a negative charge allowed to spread out over a larger volume is a more stable negative charge. In Lesson 5, we learned that both positive and negative charges can be spread out (delocalized) through resonance. The same idea applies to considering acid strength.

As established above, if the conjugate base anion is more stable, the acid producing that anion is a stronger acid. We can conclude that, all else being equal, the stronger acid in a pair of molecules will be the one that has more resonance delocalization of the negative charge on the conjugate base anion. Let's compare water and acetic acid:

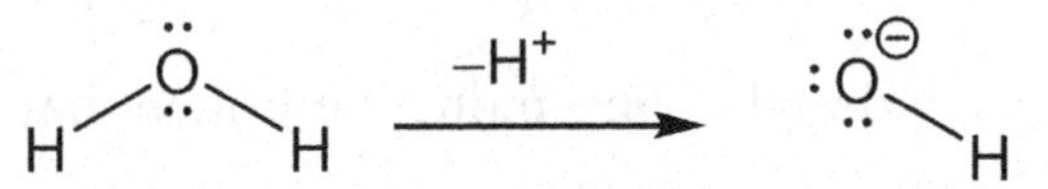

The negative charge resides on an O-atom, for both OH^- and CH_3COO^-. However, the CH_3COO^- is resonance stabilized, as shown, and it appears the negative charge is being carried by two O-atoms on the CH_3COO^- ion, and certainly more stable (more diffuse).

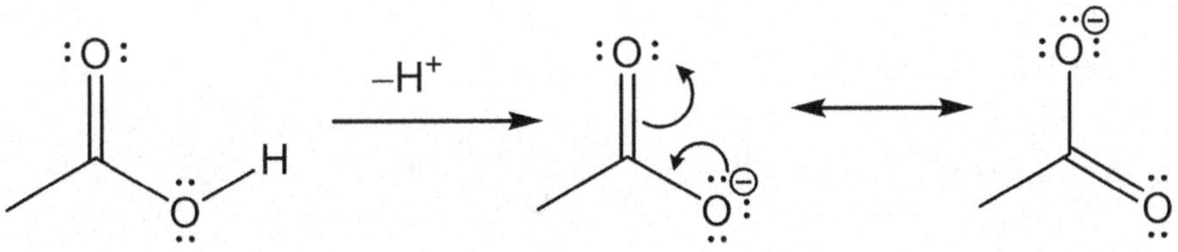

Inductive effect

Inductive effects can be divided into two types: *stabilizing* and *destabilizing*. In the context of anion stability, we need a group *near* the anionic atom in the conjugate base with some attraction for the negative charge in order to *induce* some stabilization. In other words, we need an electron-withdrawing group, namely an electronegative atom or group.

Conversely, *destabilizing inductive effects* are observed when a nearby group has a repulsive interaction with the anionic atom's negative charge. An example of a stabilizing inductive effect is an attractive force between the anionic atom and a nearby partial positive end of a polar bond:

CH_3COOH $Cl-CH_2COOH$ CCl_3COOH

$pK_a = 4.75$ $pK_a = 2.81$ $pK_a = 2.81$

The electronegative Cl-atom must be responsible for the increase in acidity. An increase in the number of Cl-atoms also causes an increase in acidity.

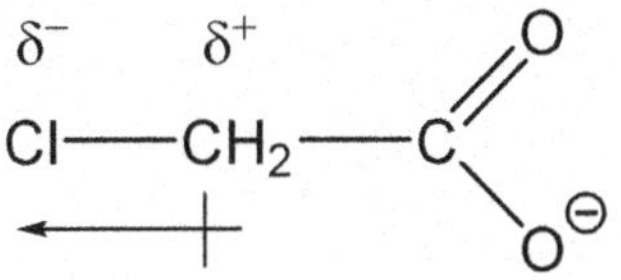

The Cl-atom is considered to be an electron-withdrawing group (EWG). Additional resonance structures cannot be legally drawn to justify the lower pKa value, so we say the negative charge is stabilized inductively through the sigma bonds. The EWG helps disperse the negative charge.

Furthermore, the magnitude of the effect, whether stabilizing or destabilizing, is greater the closer the attractive/repulsive unit is to the anionic atom.

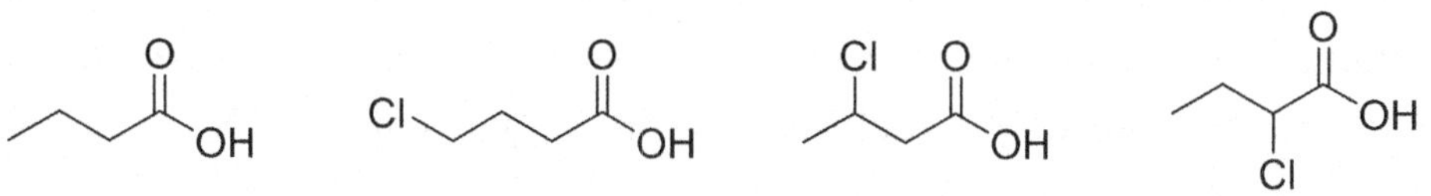

pKa = 4.8   4.5   4.5   2.9

As mentioned above, a destabilizing inductive effect can be observed. In contrast to an electron-withdrawing group (which attracts a negative charge), an inductive electron-donor will destabilize a negative charge. Alkyl groups are electron-donating groups and will destabilize the conjugate base anion, thus making the acid weaker.

It should also be noted here that these same stabilizing/destabilizing inductive effects have the opposite effect on positive charges, like the carbocations we will discuss later.

Example 7.1

Which is the stronger base for each pair?

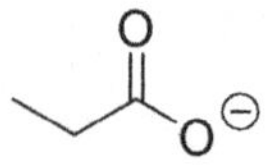

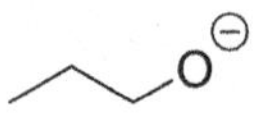

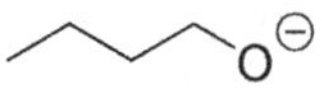

Solution 7.1

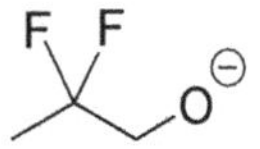

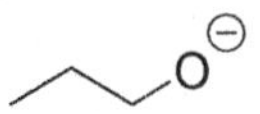

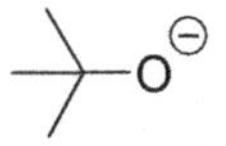

a) $CH_3O^{\ominus}$ is the stronger base because the negative charge is on the smaller O-atom.

b) 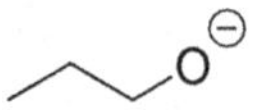 is the stronger base because the negative charge is on only one atom, not two. Resonance structures can be be drawn on the first structure, suggesting the negative charge can be shared by two oxygen atoms.

c) ![structure: CH2=CH-CH2-O⁻] is the stronger base because the negative charge is more concentrated. The two fluorine atoms on the first structure inductively diffuse the negative charge, essentially lessening the charge density on that oxygen atom.

d) ![structure: (CH3)3C-O⁻] is the stronger base because the negative charge is more concentrated. Recall that alkyl groups, like the methyl groups, donate electron density. They essentially put more electron density on the oxygen atom, making it more negatively charged and destabilized.

Lesson 7.2. Influence of Protonation State and Assessing Basicity

Protonated oxygen atoms and nitrogen atoms are more acidic than their non-protonated form.

$$CH_3OH \quad \text{vs.} \quad CH_3\overset{\oplus}{O}H_2 \qquad\qquad CH_3NH_2 \quad \text{vs.} \quad CH_3\overset{\oplus}{N}H_3$$

$$pKa = 15.5 \qquad\qquad -2.5 \qquad\qquad\qquad 40 \qquad\qquad\qquad 10.7$$

The $CH_3OH_2^+$ is similar in pK_a and structure to the hydronium ion, H_3O^+ ($pK_a = -1.7$). The $CH_3NH_3^+$ is similar in pK_a and structure the ammonium ion, NH_4^+ ($pK_a = 9.25$). The difference is that a methyl group ($-CH_3$) has replaced a H-atom.

We can also apply the same structural factors above to evaluate basicity. An obvious consideration is that a negatively charged atom is a stronger base than the same atom without a charge. For example, OH^- is a stronger base than H_2O.

Weak bases, according to our approach, have a diffuse negative charge, and the spreading of that negative charge can be accomplished through resonance, size, inductive effect, and hybridization. Conversely, a strong base would possess a concentrated negative charge. Any factor that takes electron density away from a single atom will make the base weaker.

The same ideas can be applied to neutral bases. $(CH_3)_3CNH_2$ would be expected to be a stronger base than CH_3NH_2 because the two additional methyl groups inductively donate electron density on the nitrogen atom. Let's consider some pK_b values and see if we can defend the values:

name	base	K_b	pK_b	comment, relative to ammonia
ammonia	NH_3	1.8×10^{-5}	4.75	
aniline		7.4×10^{-10}	9.13	A weaker base because the sp^2 carbon is more electronegative than an sp^3 carbon, an electron withdrawing effect. We could also draw resonance structures and push nitrogen's lone pair onto and around the benzene ring.
ethylamine	$C_2H_5NH_2$	4.5×10^{-4}	3.35	Alkyl groups inductively donate electron density. Here, the $-C_2H_5$ group is putting more electron density on the nitrogen atom, making it more partial negative, thus a little "hotter" of a base. This makes the N-atom a stronger base.

Lesson 8.1. Using pK_a to Predict Reactions

To predict the outcome of an acid-base reaction, note that the *more acidic* proton will be donated:

$$CH_3NH_2 \; + \; CH_3OH \; \rightleftharpoons \; CH_3\overset{\oplus}{N}H_3 \; + \; CH_3O^{\ominus}$$

$$pKa = 40 \qquad pKa = 15.5$$

To predict which side of the equilibrium is favored, the equilibrium always shifts from the stronger acid to the weaker acid.

reactants favored

$$\longleftarrow$$

$$CH_3NH_2 \; + \; CH_3OH \; \rightleftharpoons \; CH_3\overset{\oplus}{N}H_3 \; + \; CH_3O^{\ominus}$$

$$pKa = 15.5 \qquad pKa = 10.7$$

Lesson 8.2. Acid-base Properties of Amino Acids

Amino acids contain two (or more) functional groups – an amino group (a base) and a carboxylic acid group (an acid). As such, an amino acid can behave as an acid or a base and is said to be *amphoteric*. At the proper pH, an amino acid will assume a dipolar structure having both a positive charge and a negative charge. Such a structure is also known as a *Zwitterion*.

amino acid   Zwitterion

One of the most misunderstood ideas related to amino acids is predicting which form will predominate at a specific pH. Amino acids often have more than one acid or base present, but first we will address a simpler amino acid. For an amino acid like valine the literature will provide two pK_a values:

This is the neutral form ...   ...but the pK_a values are really for this form:

Amino acids are commonly written in their neutral form with pK_a values reported below. It is assumed that we know which pK_a goes to which functional group. As a rule of thumb, the lower pK_a value applies to the carboxylic acid and the higher pK_a value goes to the *protonated form* of the amino group.

We can control the form an amino acid assumes by manipulating the pH of a solution through the addition of a base (like NaOH) or an acid (like HCl). To predict which form of the amino acid dominates, we must consider the solution pH and each pK_a value separately and ask, "Is the proton on or off?" Let's try valine and say we added valine to water and adjusted the pH to 6.0. Which form of the amino acid would predominate?

Our adjusted solution pH of 6.0 is below 9.74 so there is enough H^+ to keep the proton on the amino group. The solution pH is below the pK_a. It's more acidic than the amino group pK_a. *Proton on.*

Our adjusted solution pH of 6.0 is above 2.29 so the solution is basic enough to remove the H^+ from the -COOH group. The solution pH is above the pK_a. It's more basic than the -COOH group pK_a. *Proton off.*

Based on the above reasoning we can conclude the following form of the amino acid would prevail at a pH of 6.0:

"Proton on" because solution pH (6.0) is below the pK_a (9.74).

"Proton off" because solution pH (6.0) is above the pK_a (2.29).

Lesson 8.3. Henderson-Hasselbalch Equation Review

In general chemistry, we encountered the Henderson-Hasselbalch equation when learning about buffers. If we could classify a question as a buffer question, then we pulled out the Henderson-Hasselbalch equation.

for the weak acid, HA: $\quad$ HA $\rightleftharpoons$ H^+ + A^- $\qquad$ $pH = pK_a + \log \dfrac{[A^-]}{[HA]}$

We could make various calculations related to this equation. Here, though, we're not all that interested in doing calculations; we should understand what this equation means in terms of what species predominate at certain pH values. Let's look at the classic example of acetic acid, CH_3COOH, with a pK_a of 4.7. According to the Henderson-Hasselbalch equation, we can prepare a solution buffering at a pH of 4.7 by having equal molar amounts of CH_3COOH and CH_3COONa.

There are two ways make this happen. One, we go to the stockroom and get an equal number of moles of CH_3COOH and CH_3COONa and add them to water. Another way is to use NaOH and neutralize exactly one-half of the CH_3COOH; this also gives us equimolar amounts of CH_3COOH and

CH3COONa. If we have a perfect 1:1 molar ratio of CH_3COOH and CH_3COONa then the solution pH will be 4.7. The H–H equation is reduced to $pH = pK_a$ because log1 equals 0.

What if we adjusted the pH of the solution to 5.7? What does this imply?

If $\quad 5.7 = 4.7 + \log \dfrac{[A^-]}{[HA]}$, this means that $\log \dfrac{[A^-]}{[HA]}$ equals 1.0, which says the $\dfrac{[A^-]}{[HA]}$ ratio is 10.

Correct? This implies that at a pH of 5.7 we have a 10:1 ratio of CH_3COO^- to CH_3COOH, which is 10 times more of the acetate ion than the acetic acid molecule. At a pH of 6.7 we would have 100:1 ratio of CH_3COO^- to CH_3COOH ... 100 times more acetate ion than acetic acid! At a pH of 7.7 we would have some 1000 times more CH_3COO^- than CH_3COOH.

If we take the pH below the ideal 4.7 pH, we can deduce that for every pH unit below 4.7 we will have 10 times more of the weak acid than the conjugate base. In other words, a pH of 2.7 implies that there are 100 acetic acid molecules for every one acetate ion present. This kind of reasoning is useful in quantifying just how predominant one species is over another when a specific pH is reached.

Lesson 9. Constitutional (Structural) Isomers and degree of Unsaturation

There are different ways that the same set of carbon atoms can be arranged for a given molecular formula. As an example, a molecule with the formula C_5H_{12} could be any of these:

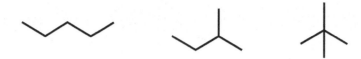

If two molecules have identical molecular formulae but differ in their bond connectivity like this, we term these **constitutional isomers** or **structural isomers**. Notice that in the structures above all of the C–C bonds are single bonds and there are no rings. In this case, the C atoms are **saturated** with as many H atoms as possible. In order to make a multiple bond between C or to make a ring, some H atoms have to be taken off of the C. For example, if we want to make a compound having a double bond or a five-membered ring, the formula will be C_5H_{10}:

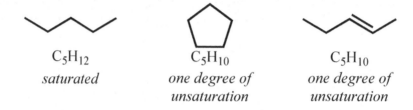

C_5H_{12}

saturated

C_5H_{10}

*one degree of
unsaturation*

C_5H_{10}

*one degree of
unsaturation*

Notice that, compared to the saturated alkane having the same carbon count, each π bond or ring that we add requires us to remove two H atoms, or one H_2 molecule, compared to the saturated structure. For this reason, each ring or π bond in a structure is referred to as a **degree of unsaturation** (DU). By looking at a structure it is relatively easy to determine the degree of unsaturation, as illustrated by Example 9.1.

Example 9.1

How many degrees of unsaturation are present in a molecule of Lexapro:

Lexapro

Solution 9.1

We count up a total of eight π bonds and three rings, for a total of eleven degrees of unsaturation.

42

It is also possible to determine the DU for a molecule given only its formula. A general formula for this is:

$$\text{Degree of unsaturation} = \frac{2(\# \text{ of } C) + 2 - (\# \text{ of } H) - (\# \text{of } X) + (\# \text{ of } N)}{2}$$

Where "# of X" is the total number of all halogens. Note that if any of the atom types are not found in the formula, those terms will just be zero in the formula. Another notable point is that if there are O atoms in the formula, their presence does not influence the DU, so "# of O" does not show up in the equation.

Example 9.2

How many degrees of unsaturation are present in a molecule having each of the following formulae:

a. $C_{13}H_{16}BrCl$

b. C_6H_7N

c. $C_{13}H_{15}NO$

Solution 9.2

a. Using the formula, we get: DU = [(2*13) + 2 – 16 – 2 + 0]/2 = 5
b. Using the formula, we get: DU = [(2*6) + 2 – 7 – 0 + 1]/2 = 4
c. Using the formula, we get: DU = [(2*13) + 2 – 15 – 0 + 1]/2= 7

The ability to determine the degrees of unsaturation and to recognize that numerous structural isomers may be possible for a given formula leads us to develop a set of simple steps to help us draw structural isomers from a given formula:

Step 1: Calculate the degree of unsaturation then draw the <u>main chain</u> (all carbons connected to each other in one chain).

Step 2: Draw the <u>main chain minus 1 carbon</u>. Add the CH_3 to as many positions as possible. **Never** add it to the ends of the parent chain.

Step 3: Draw the <u>main chain minus 2 carbons</u>. Add two one-carbon groups or one two-carbon group to as many positions as possible. Never add them to the ends of the carbon. Try not to repeat same structure.

One helpful hint as you work through this process is first to number each parent chain as you draw it to avoid getting confused later on. Then, when adding the branches, make sure you do not add a branch having n carbons to the nth carbon in the parent chain; otherwise you will change the parent chain to a longer one and end up reproducing a structure you have found from a prior step.

Example 9.3

Draw all of the structural isomers having the formula C_6H_{14}.

Solution 9.3

If we employ Step 1 and calculate the degree of unsaturation, we find that degree of unsaturation is zero. So, we have one isomer like this:

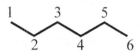

Next, employ Step 2, drawing a five-carbon main chain and adding the CH_3 to as many positions as possible. This produces two more isomers:

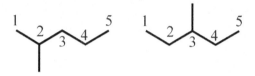

Finally, employ Step 3, drawing a main chain of four carbons in length and adding two CH_3 groups or one C_2H_5 group to as many positions as possible. This gives us the final two possibilities:

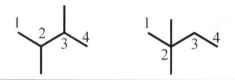

Lesson 10. Naming Alkanes and Alkyl Halides

Lesson 10.1. Naming Linear Alkanes and Substituents

In Lesson 4.2, we saw examples of the functional groups typically covered in introductory organic chemistry. Alkanes are molecules consisting of only carbon and hydrogen atoms, with all the carbons interconnected by only single bonds. As we saw in our discussion of structural isomers, a formula alone is not enough to tell us the exact molecule described. It is therefore important to develop a naming system (a nomenclature convention) for alkanes to use in our studies. **Linear alkanes** have all carbons attached in a linear chain, with no C-containing branches, called **substituents**, coming off of the chain other than hydrogen. Below are shown the linear alkanes up to ten carbon atoms in length along with their names. On the right is the name for substituent chains having the same number of C atoms as the corresponding linear alkane on the left. In addition to these alkyl substituents, halide substituents (fluoro for -F, chloro for Cl-, bromo for Br- and iodo for I-) can be used in **naming alkyl halides** following the same rules in we will learn here for alkanes.

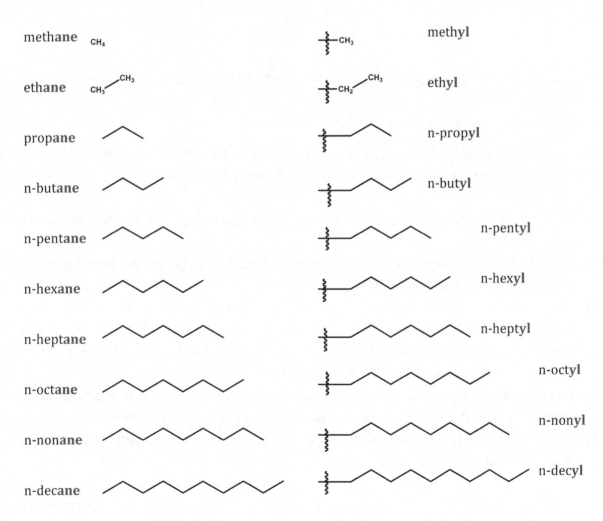

Note that some of the linear chains have an *n-* in front of their names. This *n-* stands for "normal", and it is sometimes left off of the name and we assume that it is a "normal" linear chain. On the other hand, some arrangement of hydrocarbon chains that are placed as substituents have specific names that use other designators in front of the part of the word telling us how many C are present. The more common ones are provided below for reference.

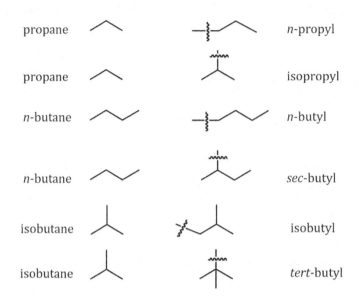

propane			n-propyl
propane			isopropyl
n-butane			n-butyl
n-butane			sec-butyl
isobutane			isobutyl
isobutane			tert-butyl

Lesson 10.2. Naming Branched Alkanes and Alkyl Halides

We will now use a series of examples to illustrate how to use the following six-step guide for naming alkanes. Note that, for most alkanes, you will only need some of these rules.

1) Find the longest chain. This is the 'parent chain'; the other things coming off of the parent chain are its substituents.
2) Number the carbon atoms in the 'parent chain' in the way that gives the lowest number to the substituent closest to an end of the parent chain.
3) If more than one type of substituent is present, name in alphabetical order.
4) If more than one of the same substituent are present on your parent chain, use di, tri, tetra, etc., prefixes to denote this (these prefixes do not count when alphabetizing, though; neither do the *n-*, *sec-*, or *tert-* prefixes; however, the *iso* prefix DOES count)
5) If numbering leads to the same lowest number substituent in either direction, the correct numbering gives the lowest number to the substituent that is first alphabetically.
6) If you find two different possible parent chains of the same length, you choose the one with more substituents coming off of it.

Consider the molecule on the left. The longest chain (step 1) has five carbons, so the parent chain is pentane (highlighted on the right).

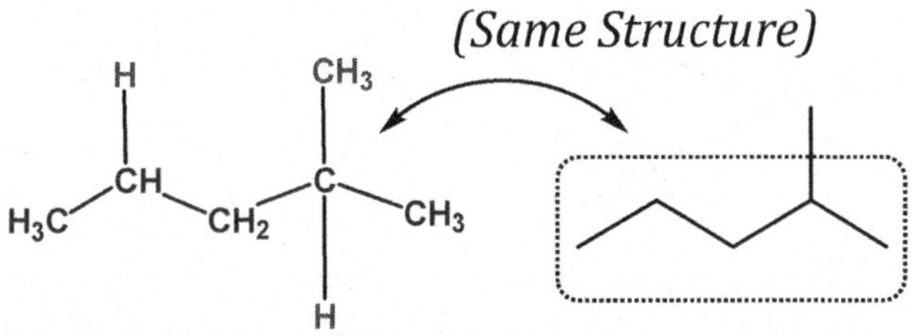

(Same Structure)

Having identified the parent chain, we move on to step 2, numbering the carbon atoms in the parent chain such that the substituent (here a methyl) has the lowest number:

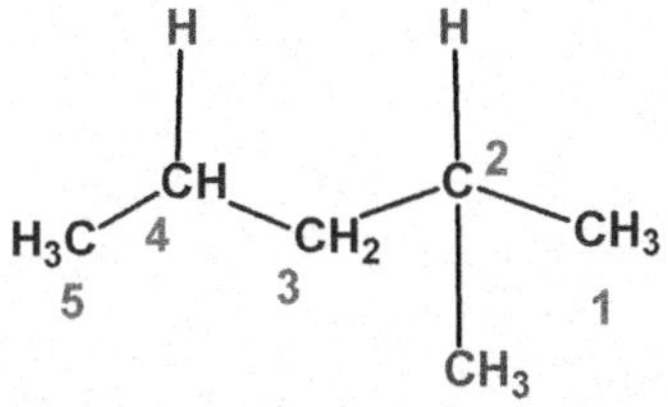

This molecule is thus called **2-methylpentane**. Note that there is always a dash between a numeral and the substituent to which it refers. Consider another example, in which we have already applied rules 1 and 2, and have found a heptane chain having a methyl substituent at position 2 and an ethyl substituent at position 4:

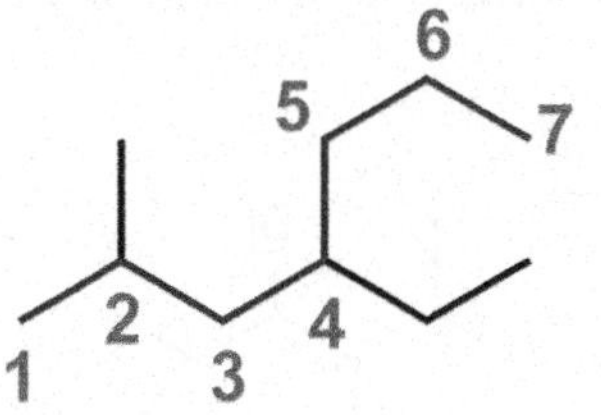

We now need to apply rule 3 (list substituents alphabetically) to compose the final name of **4-ethyl-2-methylheptane**.

In a case in which a parent is substituted by more than one of the same substituent, for example:

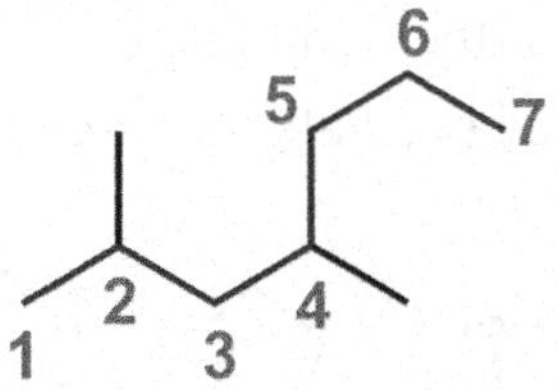

We will need rule 4 involving use of di-, tri-, etc. prefixes. The molecule above is thus properly called **2,4-dimethylheptane**.

If we encounter a case like this:

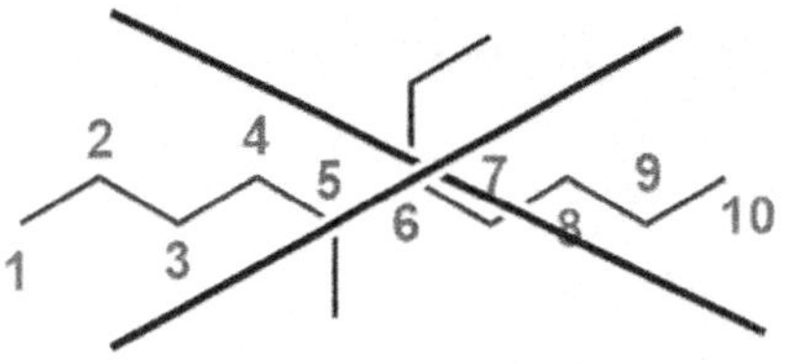

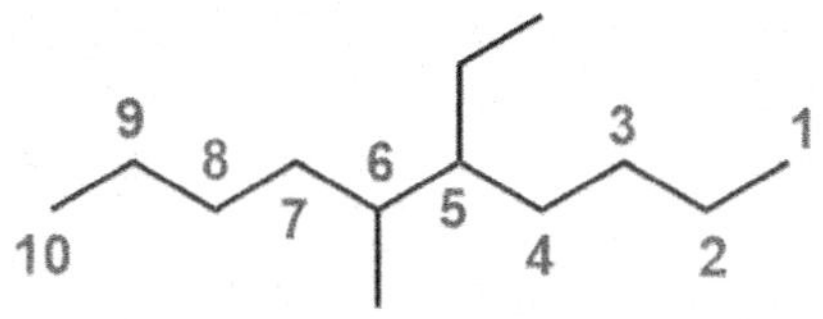

In which there are two possible ways to number the parent chain, we need to apply the fifth rule. This rule tells us that in a tie, give the alphabetically first substituent the lower number, as shown on the right, above. This approach allows us to properly name the molecule as **5-ethyl-6-methyldecane**.

We may encounter a molecule in which two possible longest parent chains are identified, as in the example below.

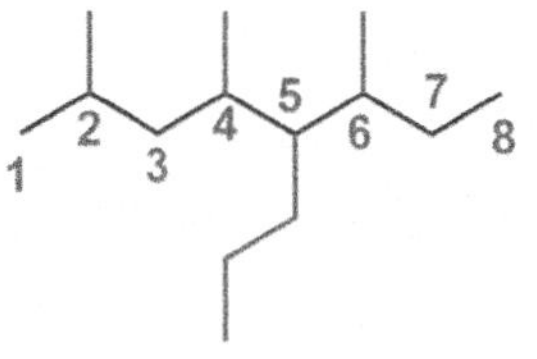

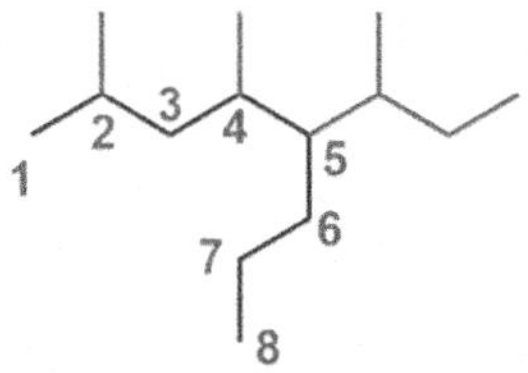

In these cases, rule 6 tells us to choose the parent chain having more substituents (the one on the left), so this molecule is properly called **2,4,6-trimethyl-5-propyloctane**.

Alkyl halides are named using the same set of rules as are alkanes, as illustrated by the example below:

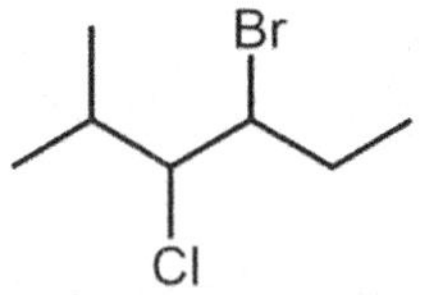

4-bromo-3-chloro-2-methylhexane

Lesson 10.3. Naming Substituted Cycloalkanes

Alkanes that have a cyclic structure, rather than linear, are called **cycloalkanes**. The structures and names for the most common unsubstituted cycloalkanes are provided below.

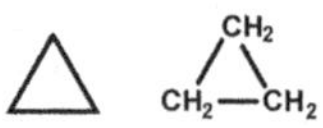

cyclopropane 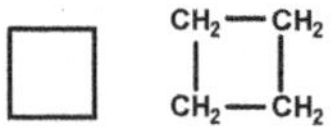   cyclobutane 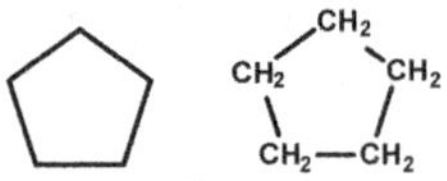   cyclopentane 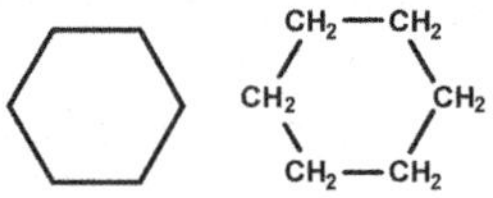   cyclohexane

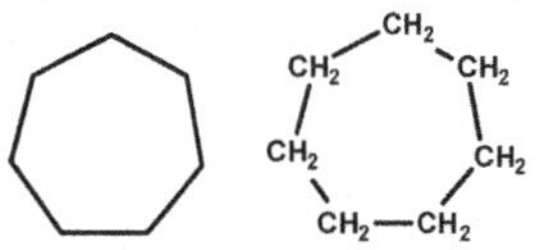

cycloheptane 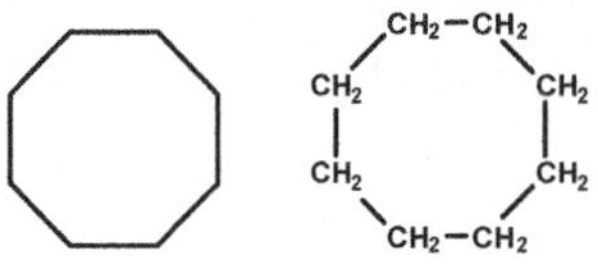   cyclooctane

A cycloalkane may be substituted just as a linear alkane may be substituted. The cycloalkane can also be the parent chain, if it is the longest contiguous chain of carbon atoms in the molecule. A parent chain may be composed *either* of C atoms in a cycloalkane *or* C atoms in a linear chain. One may not mix C atoms from both types of alkanes to form a parent chain, so:

4-cyclopentyl-2,3-dimethylhexane   1-ethylcyclohexane

The rules for naming linear alkanes also apply to cycloalkanes.

Lesson 11. Naming Saturated *O*- and *N*-Containing Molecules

Lesson 11.1. Types of Carbon Atoms

In discussing organic molecules, it is often useful to discuss functional groups in terms of what type of carbon is attached to the functional group. Carbons can be referred to a primary (1°), secondary (2°) or tertiary (3°).

A "primary" carbon is bonded to **one carbon**. Hydrogens on this carbon are "primary".

A secondary carbon is bonded to **two carbons**. Hydrogens on this carbon are "secondary".

A tertiary carbon is bonded to **three carbons**. Hydrogens on this carbon are "tertiary".

A quaternary C is bonded to **four carbons**. These C have no hydrogens.

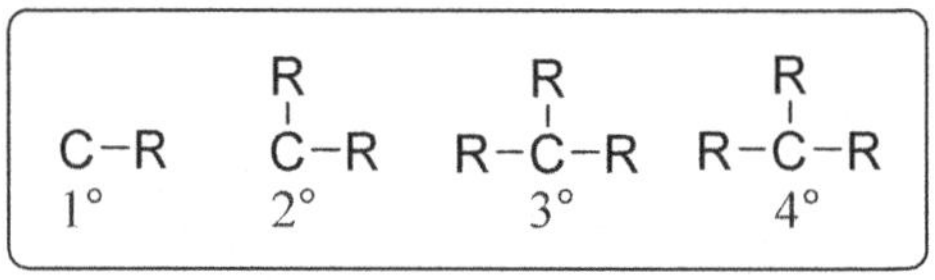

Example 11.1

Label each carbon as 1°, 2°, or 3°

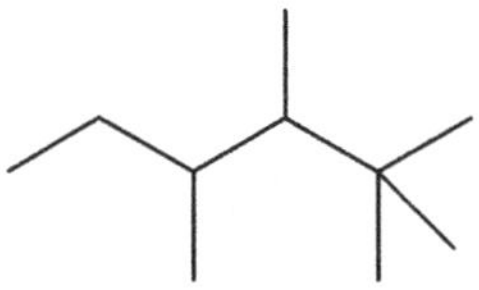

Solution 11.1

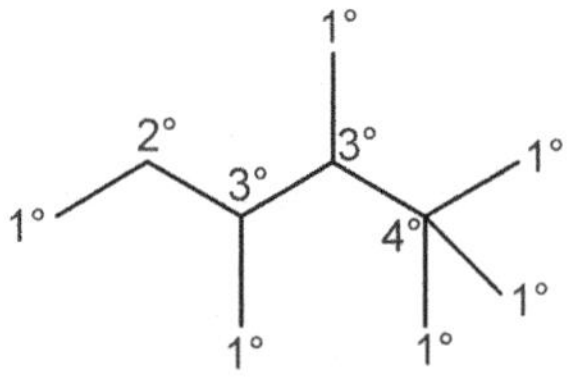

The types of Alkyl Halides (R—X) are also subdivided based on what type of carbon is attached to the halogen (X = F, Cl, Br or I). A primary alkyl halide as the halogen on a primary C, etc.:

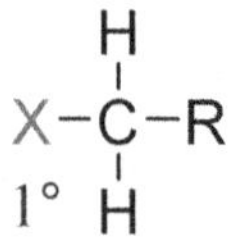

primary alkyl halide

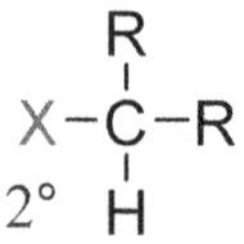

secondary alkyl halide

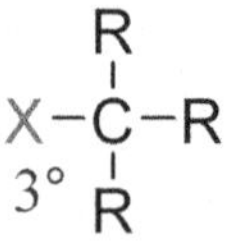

tertiary alkyl halide

Lesson 11.2. Naming Alcohols

A few common alcohols have common names of the form "alkyl alcohol". These common names are mostly used for methyl alcohol (CH_3OH), ethyl alcohol (CH_3CH_2OH). For systematically naming alcohols, we use the rules for naming alkanes and cycloalkanes as a starting point, with the following adjustments: Replace the "e" at the end of the alkane name with "ol".

The alcohol is always given the lowest possible number. Note that this means that the alcohol is always given the "1" position in cycloalkanes (so there is no need to add a number, because it is always 1).

Place the number indicating the position of the alcohol directly before the parent chain name (which now ends in "ol").

Again, we will illustrate how the rules are applied with examples:

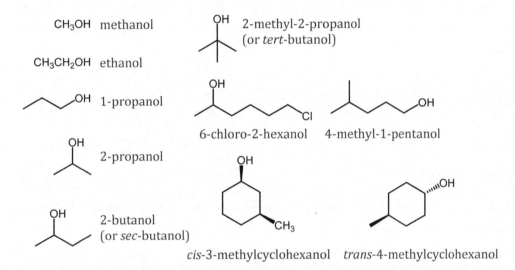

Just as we saw for alkyl halides, alcohols can be referred to as primary (1°), secondary (2°) or tertiary (3°) depending on the carbon attached to the hydroxyl (–OH) group:

$$\underset{\text{primary alcohol}}{\overset{\overset{\displaystyle H}{|}}{HO-\underset{\underset{\displaystyle H}{|}}{C}-R}\ 1°} \qquad \underset{\text{secondary alcohol}}{\overset{\overset{\displaystyle R}{|}}{HO-\underset{\underset{\displaystyle H}{|}}{C}-R}\ 2°} \qquad \underset{\text{tertiary alcohol}}{\overset{\overset{\displaystyle R}{|}}{HO-\underset{\underset{\displaystyle R}{|}}{C}-R}\ 3°}$$

Lesson 11.3. Naming Ethers

When we name ethers, we divide them into two types: **symmetrical** (the two R's are same, for example $CH_3CH_2OCH_2CH_3$), and **unsymmetrical** (two R's are different, for example $CH_3OC_2H_5$).

We can effectively name ethers with either **common names** or **systematic names**. Common names are formulated by first naming each R group (the R and R^1 groups in the diagram below):

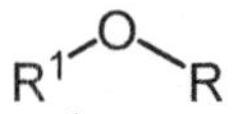

$$R^1 \diagdown O \diagdown R$$
Alkyl (R^1) alkyl ether

The two alkyl groups are then listed alphabetically, followed by "ether". If the ether is symmetrical, it is simply named "di*alkyl* ether." These examples illustrate the common naming method:

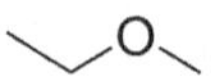

ethyl methyl ether 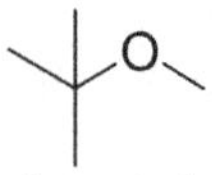   *t-butyl methyl ether* 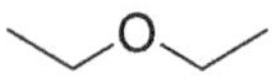   *diethyl ether*

IUPAC systematic names are often used for ethers as well. In this approach, the longer alkyl group is named as the parent chain and the O with the shorter alkyl group is named as an **alkoxy** substituent (methoxy, ethoxy, or alkyloxy for longer C chains). These two examples illustrate the systematic naming method:

1-ethyoxybutane 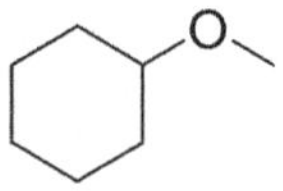   *methoxycyclohexane*

Lesson 11.4. Naming Amines

For categorizing amine types, it is the number of C attached to the **nitrogen** that determines the type of amine:

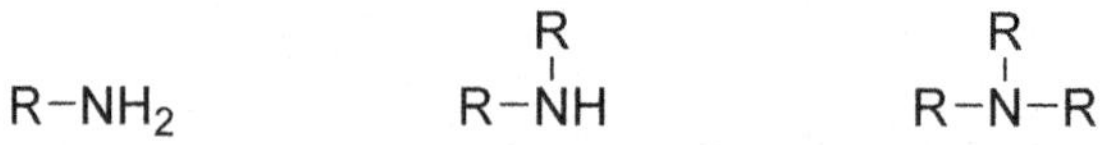

$$R-NH_2 \qquad R-\overset{\displaystyle R}{\overset{|}{N}H} \qquad R-\overset{\displaystyle R}{\overset{|}{N}}-R$$
1^0 \qquad 2^0 \qquad 3^0
primary amine   secondary amine   tertiary amine

For common naming of amines, substituents are placed in alphabetical order followed by "amine," in a fashion similar to what we saw for ethers. Amine common names all written as one word, as illustrated by these examples:

CH_3NH_2

methylamine   *methylpropylamine* 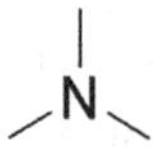   *trimethylamine* 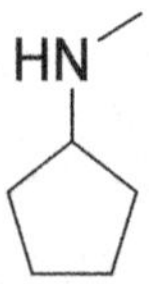   *cyclopentylmethylamine*

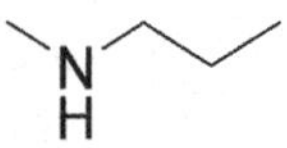

For IUPAC naming of amines, the following rules apply:
1. The main chain **must contain** the amine (C–N)
2. The parent chain is the longest to which N is attached and is named by replacing the ending "**e**" with "**amine**".
3. Numbering begins from the carbon closest to the amine.
4. The number is either given before the parent name or in-between the parent name part and the word amine.
5. A substituent on the carbon chain is indicated by a number and a substituent on the nitrogen is indicated by an "*N*."

Systematic naming of amines is illustrated by this example:

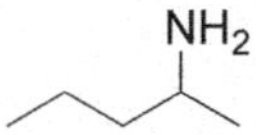

NH_2 2-pentanamine
or
pentan-2-amine

Alcohols take priority over amines. In a molecule that has both an alcohol unit and an amino unit, the amine is considered substituent, and is listed as "**amino**", as illustrated in this example:

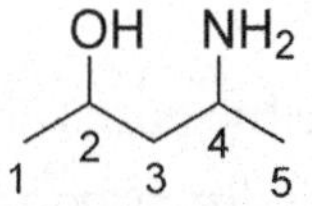

OH NH_2

4-amino-2-pentanol

When there are four groups bonded to the N, the groups are listed in alphabetical order followed by "**ammonium**" and then the name of the **counteranion**:

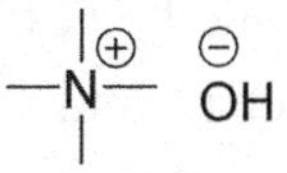

tetramethyl ammonium hydroxide

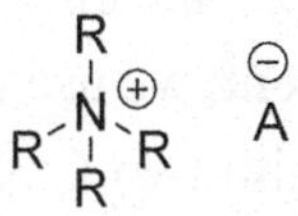

ammonium salt

Lesson 12. Naming Alkenes and Introduction to Geometrical Isomers

Lesson 12.1. Types of alkenes

We have seen that carbon atoms in a structure, as well as functional groups having heteroatoms, can be classified as being primary, secondary or tertiary. Alkenes are classified instead by how many C are bound to the C=C carbons.

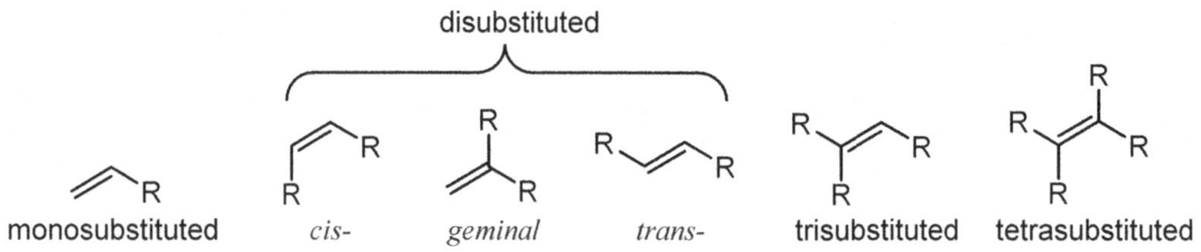

We will discuss the meaning of the *cis-*, *trans-* and *gem-* categories of the disubstituted alkenes in this lesson. In future lessons, we will also learn that the relative stability of alkenes varies from least stable (monosubstituted) on the left to most stable (tetrasubstituted) on the right.

Lesson 12.2. The "-ene" Suffix

We know how to name alkanes, and we will now use the core set of nomenclature rules from Lesson 10 as the basis to name alkenes as well. An alkene is a hydrocarbon with a C=C bond (an unsaturation). Recall that the double bond is made up of a σ bond and a π bond, a fact that will be important to remember especially when we begin learning about alkene reactivity.

When naming alkenes, we replace the "-ane" suffix with "-ene". So, if we have a 6-carbon chain with one C=C bond, it is a "hexene"; an 8-carbon chain with one C=C bond is an "octene". If two or three C=C bonds are present, we use the suffix "-diene" or "-triene", and we add an "a" to the end of the root. A 6-carbon chain with two C=C bonds is a "hexadiene", and an 8-carbon chain with three C=C bonds is an "octatriene".

Lesson 12.3 Alkene Priority in Hydrocarbons

We first pick as parent the longest chain that contains the C=C bond. If multiple C=C bonds are present, we select as parent the chain that contains the greatest number of C=C bonds (these parent chain rules supersede the ones for alkanes). We then number the parent chain to give the C=C bond(s) the lowest possible substituent numbers.

<u>Example 12.1</u>

Provide the unambiguous systematic name for the molecule shown below:

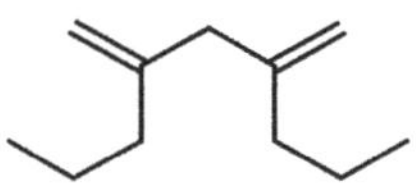

Solution 12.1

Even though the longest chain present in this molecule has 9 carbons, we have to select as parent the chain that has the greatest number of C=C bonds. For this molecule, it is a 5-carbon chain with 2 C=C bonds. It does not matter if we start numbering left to right or right to left in this example, because the molecule is symmetric. The C=C bonds begin at carbons 1 and 4, so the parent is a 1,4-pentadiene. Two substituents are present at carbons 2 and 4. We would need to use the complex substituent rules to assign them names, and they are both propyl substituents. The complete systematic name is thus 2,4-dipropyl-1,4-pentadiene.

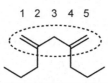

Lesson 12.4. Alkene Priority in Naming

If an alkene also contains an OH group, then that OH group takes priority over the C=C bond. Thus, we must select as parent the chain that contains an OH group, even if that results in fewer C=C bonds being included. Similarly, we must number the parent chain to give the OH group the lowest possible substituent number, regardless of what substituent numbers that assigns to the C=C bonds. To assign a name to the molecule, the "ene" suffix becomes "en" and is listed before the "ol" suffix which ends the name. A 6-carbon chain with one C=C bond and one OH group would thus be a "hexenol". For a complex molecule such as this, the substituent numbers can be placed immediately before the suffixes to enhance clarity.

Example 12.2

Provide the unambiguous systematic name for the molecule shown below:

OH

Solution 12.2

The longest chain in this molecule has 7 carbons, but we must pick the longest one that has both the OH group and the C=C bond. The longest chain that meets these criteria has only 4 carbons, so the root is "butenol". We have to number from left to right because that gives the OH group a substituent number of 2 (going from right to left it would have a 3). As a result, the C=C bond begins at carbon 3, so the parent molecule is "3-buten-2-ol". We next need to identify the substituents and their positions in this molecule, which are methyl at carbon 2 and an *n*-butyl at carbon 3. The complete systematic name is thus "3-*n*-butyl-2-methyl-3-buten-2-ol".

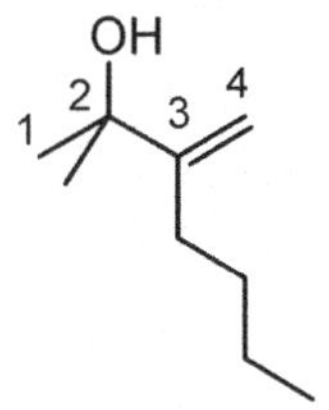

Lesson 12.5 *The cis-/trans- Prefix Convention*

The two isomers of 3-hexene shown below differ only by whether the two ethyl substituents are on the same side or opposite sides of the C=C bond:

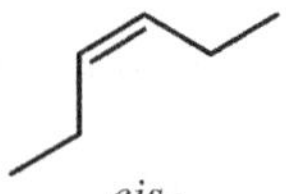

cis- 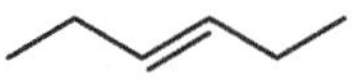   *trans-*

Rephrasing this in more general terms, the two isomers differ only by the spatial orientation of their substituents, and the two isomers cannot be interconverted without breaking any bonds. These types of situations lead to the two being classified as configurational isomers (**geometric isomers**), a class of stereoisomers. In these cases, the C=C bond is said to be stereogenic. Note, however, that it is only when each carbon in a C=C bond has two non-identical substituents that two geometric isomers are possible; if two of the substituents on one C atom are identical, then switching them will not yield a different stereoisomer.

The simplest type of stereogenic C=C bond is one in which each carbon in the C=C bond has one hydrogen and one non-hydrogen substituent, and the term for this type of molecule is a disubstituted, internal alkene. In contrast, a disubstituted, external alkene is a molecule in which one carbon in the C=C bond has two hydrogen substituents and the other carbon and has two non-hydrogen substituents.

If we consider 2-methyl-1-pentene (see below), we can see that switching the methyl and *n*-propyl substituents does not produce a new molecule, so 2-methyl-1-pentene does not have multiple stereoisomers. For similar reasons, a monosubstituted alkene like 1-hexene (another constitutional isomer of 3-hexene) does not have multiple stereoisomers.

2-methyl-1-pentene

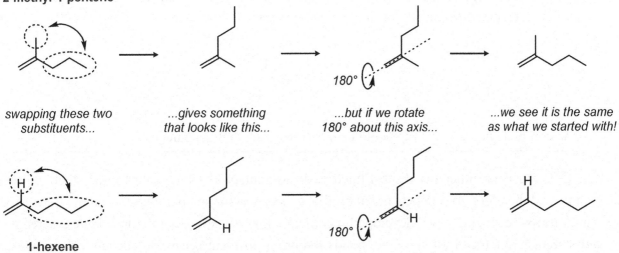

swapping these two substituents...	*...gives something that looks like this...*	*...but if we rotate 180° about this axis...*	*...we see it is the same as what we started with!*

1-hexene

For an internal disubstituted alkene, if both substituents are on the same side of the line passing through the C=C bond, we apply the prefix "*cis*-". If the substituents are on opposite sides of the line passing through the C=C bond, we apply the prefix "*trans*-". Consider again the two configurational isomers of 3-hexene as an example:

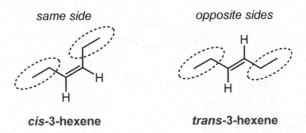

cis-3-hexene   *trans*-3-hexene

Note that the "*cis*-" and "*trans*-" prefixes belong to a convention developed specifically for disubstituted, internal alkenes and they can only be used for disubstituted, internal alkenes and cycloalkanes (as we will see later in the text).

Example 12.3

Provide the appropriate name for the molecule shown below:

Solution 12.3

The alkene carbon on the left has one non-hydrogen substituent, a methyl group. The carbon on the right also has one non-hydrogen substituent, an isopropyl group. Because these two non-hydrogen substituents are on opposite sides of the C=C bond with respect to each other, the proper stereochemical prefix for this molecule would be "*trans*-".

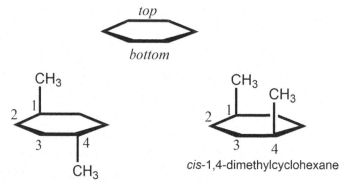

The complete name of the molecule is thus *trans*-4-methyl-2-pentene.

If a C=C bond is either (a) disubstituted and external or (b) monosubstituted, then it is not stereogenic and therefore no stereochemical prefix is necessary for an unambiguous name. Because the "*cis*-" and "*trans*-" convention was developed specifically for disubstituted, internal alkenes, stereogenic tri- and stereogenic tetrasubstituted C=C bonds require a different set of stereochemical prefixes, as described in the next section.

Lesson 12.6. Geometric Isomers of Cycloalkanes

In addition to what we learned previously for naming cycloalkanes, there is one additional consideration. A ring of atoms has a "top" face and a "bottom" face, as indicated in the picture below.

top

bottom

CH₃

trans-1,4-dimethylcyclohexane

cis-1,4-dimethylcyclohexane

The two possible ways to place the methyl groups lead to two geometric isomers. Because there are two faces of the ring, we will need to provide a prefix in front of the name of a disubstituted cycloalkane to indicate whether one group is oriented a face opposite of the other (we use the *trans*- prefix for this) or if the two groups are oriented on the same face (we use the *cis*- prefix in such cases). The two examples above illustrate this convention. The *trans*- and *cis*- forms of a disubstituted cycloalkane are examples of configurational isomers. Remember that configurational isomers have the same constitution (all atoms are attached to the same atoms in both cases) and differ only in directions that the groups point off of the structure. Configurational isomers also cannot interconvert without σ-bond breakage: if we wanted to change the *cis*- into the *trans*- isomer, we would have to break the σ-bond to the methyl substituent and switch its position, for example.

Lesson 12.7. Tri- and Tetrasubstituted Alkenes and the E-/Z- Prefix Convention

When an alkene is tri- or tetrasubstituted, the *cis-/trans-* convention is inapplicable, but its C=C bond may still be stereogenic, so we need to have a way to describe the necessary stereochemical information in that molecule's name. For example, if we add a methyl substituent to the 3-position of 3-hexene, the C=C bond is still stereogenic: in one stereoisomer, the two ethyl substituents are on the same side of the C=C bond, and in the other stereoisomer, the two ethyl substituents are on opposite sides.

The terms "*zusammen*" (German: "together") and "*entgegen*" (German: "opposed") are used to describe stereoisomers of tri- and tetrasubstituted alkenes. The way that "*zusammen*" (abbreviated as the prefix "*Z*-") and "*entgegen*" (abbreviated as the prefix "*E*-") are used is to assign priority to the substituents the two substituents on each carbon in a C=C bond (i.e., each carbon has a "higher" and a "lower" priority substituent), and then we compare the relative spatial orientation of the two "higher" priority substituents. If the two "higher" priority substituents are on the same side of the C=C bond, then we say they are "together", or "*zusammen*", and we use the prefix "*Z*-". If the two "higher" priority substituents are on opposite sides of the C=C bond, then we say they are "opposed", or "*entgegen*", and we use the prefix "*E*-". So, how do we "assign priority" to the substituents? We use a convention called the Cahn-Ingold-Prelog rules:

To name chiral molecules, then, we must first learn the convention for assigning priority to the substituents coming off a stereogenic atom, which is called the **Cahn-Ingold-Prelog** (CIP) rules. Consider this hypothetical structure to illustrate how these rules work:

Cahn-Ingold-Prelog (CIP) rules:
1) First look at the atoms directly attached to the double-bonded C. Higher atomic number = higher priority (A, B, C and D in the figure above).
2) If same atomic number, higher mass = higher priority (Deuterium > H, $^{13}C > {}^{12}C$, etc.)
3) If atoms A and B are identical, move to highest priority atom attached to A and B until a break in the tie is found (first compare A1 to B1 priority. If tie, compare A2 to B2. If tie, compare A3 to B3, etc.)

Example 12.4

Provide the appropriate stereochemical prefix for the molecule shown below:

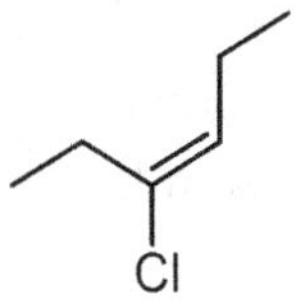

Solution 12.4

The left carbon in the C=C bond has an ethyl substituent and a chlorine atom, and the Cl has higher priority. The right carbon in the C=C bond has an ethyl substituent and a hydrogen atom, and the Et has higher priority. Because the two "hi" substituents are on opposite sides of the C=C bond, this molecule would require a stereochemical prefix of "(*E*)-".

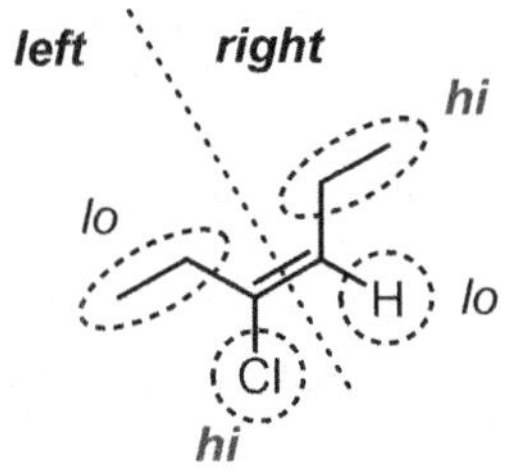

Lesson 12.8. Alkene Priority in Nomenclature

If we look at the structure of 2,3-dimethyl-3-penten-2-ol (below left), we can see that the C=C bond is stereogenic, so the name must include a stereochemical prefix to be unambiguous, which in this case would be "(*E*)-". We would thus use the CIP rules to assemble the core name and place the label of configuration in front of the name in parentheses.

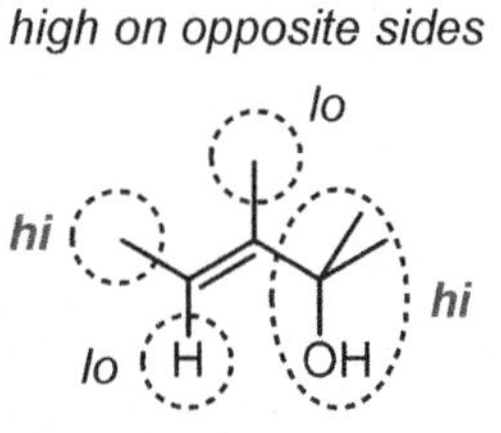

(*E*)-2,3-dimethyl-3-penten-2-ol

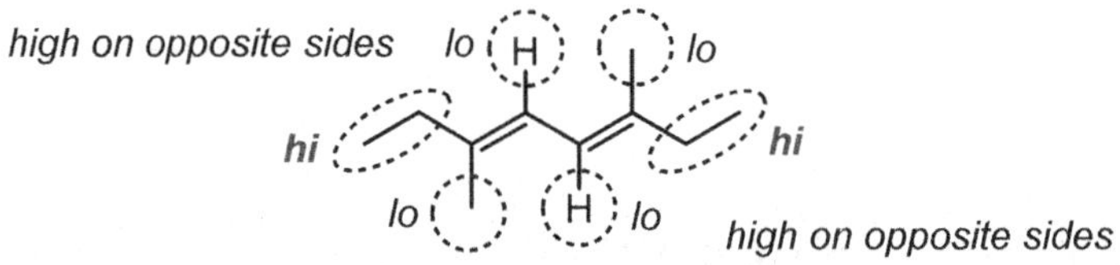

(3*E*,5*E*)-3,5-dimethyl-3,5-octadiene

What happens if a molecule has more than one stereogenic C=C bond? In this case, we would include a number with the stereochemical prefix to indicate which C=C bond has which stereochemistry. So, for the stereoisomer of 3,5-dimethyl-3,5-octadiene shown above, the complete prefix would be "(3*E*,5*E*)-".

Example 12.5

Provide the unambiguous name for the molecule shown below:

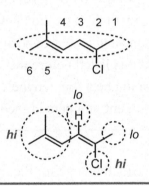

Solution 12.5

The parent of the molecule is the 6-carbon chain containing the two C=C bonds, and we number from right to left to give a lower position number to the substituent that comes earlier in the alphabet ("chloro" vs. "methyl"). Although carbons 4 and 5 are connected by a C=C bond, this C=C bond is not stereogenic because carbon 5 has two identical substituents. The double bond between carbons 2 and 3 is stereogenic: on the left carbon, the hydrogen is lower priority than the carbon chain, and on the right carbon, the methyl is lower priority than the chlorine. Because the two "hi" substituents are on the same side of the C=C bond, the prefix should be "(Z)-". The proper name for the molecule would thus be "(Z)-2-chloro-5-methyl-2,4-hexadiene".

Lesson 13.1. The "-yne" Suffix

When naming alkynes, we make use of the core nomenclature rules found in Lesson 10. For an alkyne, we replace the "-ane" suffix with "-yne". So, if we have a 5-carbon chain with one C≡C bond, it is a "pentyne"; a 9-carbon chain with one C≡C bond is a "nonyne". If two or three C≡C bonds are present, we use "-diyne" or "-triyne", and we add an "a" to the end of the root. A 5-carbon chain with two C≡C bonds is a "pentadiyne", and a 9-carbon chain with three C≡C bonds is a "nonatriyne".

Lesson 13.2. Alkyne Priority in Hydrocarbons

To determine the name for an alkyne, we first pick as parent the longest chain that contains the C≡C bond. If multiple C≡C bonds are present, we select as parent the chain that contains the greatest number of C≡C bonds. We then number the parent chain to give the C≡C bonds the lowest possible substituent numbers. If both C=C and C≡C bonds are present, we number the parent chain to give the multiple bond substituents the lowest possible numbers, regardless of whether they are C=C or C≡C bonds. If, and only if, a C=C bond and a C≡C bond would have the exact same number, do we employ the alphabetization rule, and assign priority to the C=C bond over the C≡C bond ("ene" comes before "yne" alphabetically). When both a C=C bond and a C≡C bond are present in a molecule's structure, the suffix for the C=C bond becomes "en" and it is listed before the C≡C suffix "yne". So, a 5-carbon chain with one C=C bond and one C≡C bond would be a "pentenyne". For complex molecule names such as these, it is most convenient to list the substituent numbers immediately before the suffixes.

Example 13.1

Provide an unambiguous systematic name for the molecule shown below:

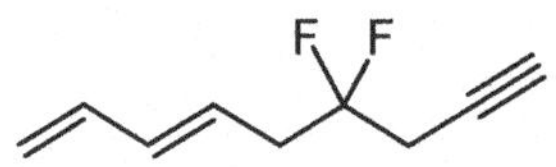

Solution 13.1

In this molecule, we can see that the parent chain contains 9 carbons, 2 C=C bonds, and 1 C≡C bond, so the base name will be "nonadienyne". We have to number from left to right to give the multiple bonds the lowest possible substituent numbers, which are 1, 3, and 8. Numbering from right to left would have given 1, 6, and 8. There are 2 fluorine substituents at carbon 6, and the C=C bond beginning at carbon 3 is stereogenic (the two non-hydrogen substituents are *trans*). The systematic name for this molecule is "*trans*-6,6-difluoronona-1,3-dien-8-yne".

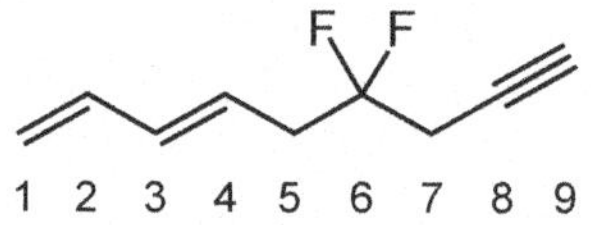

Lesson 13.3. Alkyne Priority in Heteroatom-Containing Compounds

If an alkyne also contains an OH group, then that OH group takes priority over the C≡C bond. Thus, we must select as parent the chain that contains an OH group, even if that results in fewer C≡C bonds being included. Similarly, we must number the parent chain to give the OH group the lowest possible substituent number, regardless of what substituent numbers that assigns to the C≡C bonds. To assign a name to the molecule, the "yne" suffix becomes "yn" and is listed before the "ol" suffix which ends the name. A 5-carbon chain with one C≡C bond and one OH group would thus be a "pentynol". Again, for a complex molecule such as this, we would list the substituent numbers immediately before the suffixes.

Example 13.2

Provide an unambiguous systematic name for the molecule shown below:

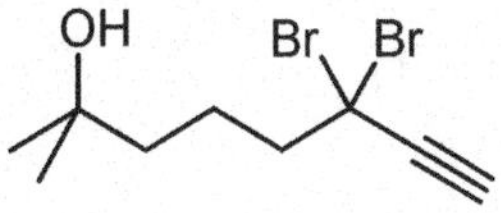

Solution 13.2

The parent chain for this molecule contains 8 carbons, 1 C≡C bond, and 1 OH group, so the base name will be "octenol". We number this chain from left to right to place the OH group at carbon 2 (going right to left would place it at 7) and the C≡C begins at carbon 7. There is 1 Me group at carbon 2, 2 Br substituents at carbon 6, and no stereocenters or stereogenic bonds. The systematic name for this molecule is "6,6-dibromo-2-methyloct-7-yn-2-ol".

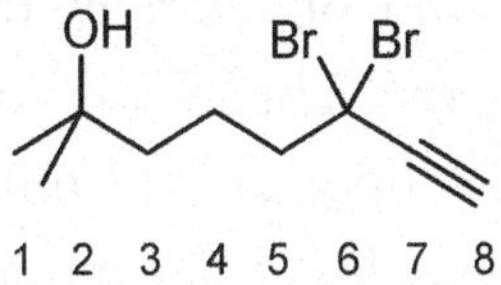

1 2 3 4 5 6 7 8

Lesson 13.4. Types of Alkynes

As we discussed in Lesson 12, a C=C bond may need a label of configuration (i.e., *cis* vs. *trans* or *E* vs. *Z*). In contrast, each carbon in a C≡C bond can have only one substituent, so no stereoisomerism is possible and thus a stereochemical prefix is unnecessary.

Alkynes can be internal or terminal, where internal alkynes can also be symmetrical or unsymmetrical depending on the two "R" groups attached to the carbons of the alkyne.

35

Lesson 14. Isomerism and Conformational Analysis I: Linear Alkanes and Newman Projections

Lesson 14.1. Newman Projections and Conformational Isomers of Ethane

Now that we have a strong foundation in analyzing attractive and repulsive forces in molecules, we can begin to evaluate how these forces can influence the geometries of molecules and the relative spatial orientation of their bonds. Consider a simple molecule like ethane. We can draw ethane in several possible ways, two of which are illustrated below:

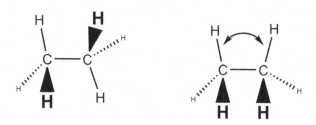

The only difference between these two representations of ethane is how the H atoms on one C are rotated relative to the H atoms on the other C. Structures that differ only by the angle of rotation about a σ-bond are called **conformational isomers**. Unlike the geometrical (configurational) isomers discussed in Lesson 12, conformational isomers can be interconverted by simple bond rotation, so we do not need to add a prefix in front of the name to distinguish these separate species (because they interconvert easily). Both of the conformations shown in the picture are simple called "ethane" as a molecular name. Note, however, that the two conformations do not have the same stability. As indicated by the double headed arrow in the structure on the right, the C–H bonds are pointed in the same way on each of the two carbons. Remembering that a bond comprises two electrons and that electrons repel one another, placing the C–H bonds close together like this will be a less stable arrangement than the conformation on the left, in which the bonds are staggered and thus farther apart from one another. This type of strain that results from twisting two bonds close to one another is called **torsional strain**. The overlapping of bonds is easier to see if we take a look at the molecule from a perspective looking down the C–C bond:

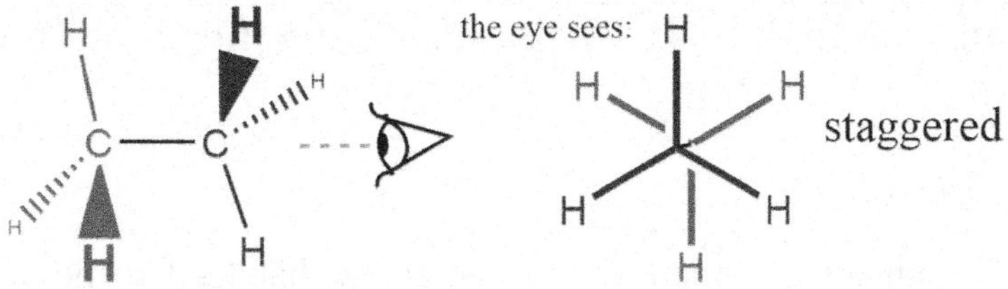

the eye sees: staggered

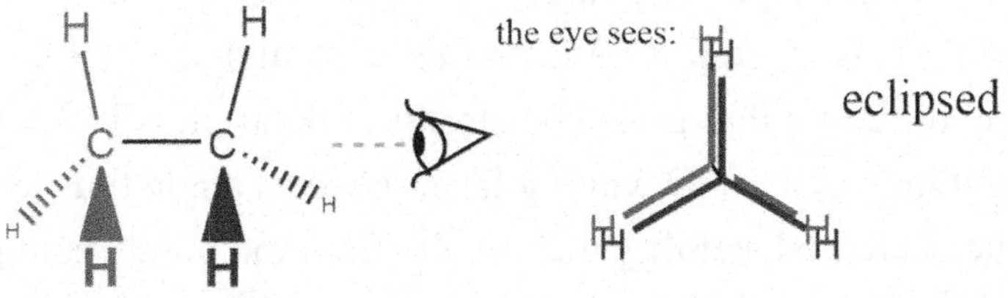

the eye sees: eclipsed

Note that it is somewhat difficult to discern which H atoms are attached to which C in the drawings on the right, so chemists use what are called **Newman Projections** for a clearer view of attachment:

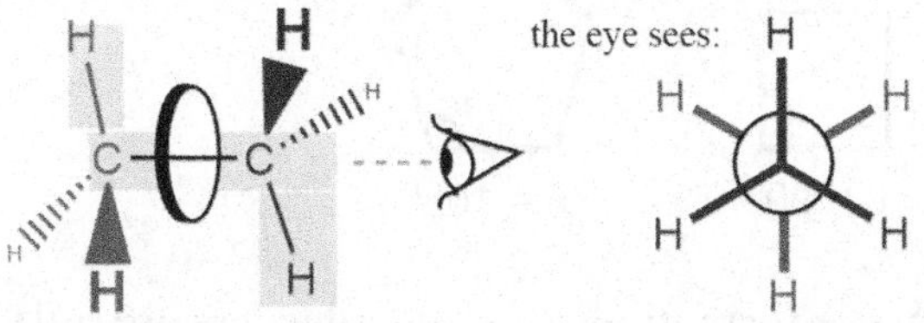

the eye sees: Newman Projection for the staggered conformation of ethane

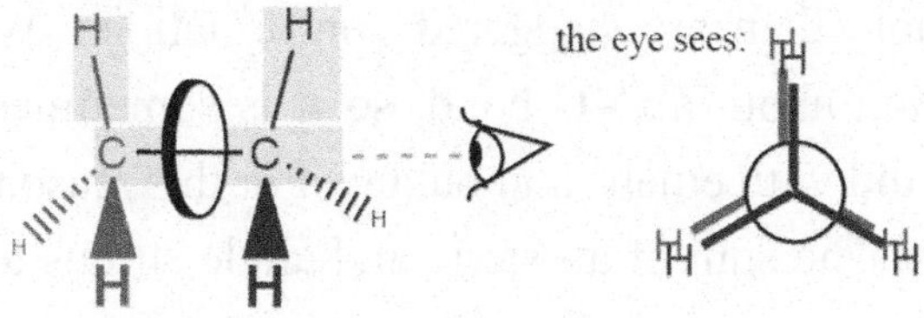

the eye sees: Newman Projection for the eclipsed conformation of ethane

We can easily see from the Newman Projections on the right that the eclipsed conformation of ethane places the electrons in the C–H bonds on the front C closer to the C–H bonding electrons on the back carbon, whereas the staggered conformation minimizes this by placing these C–H bonds as far apart as possible. A staggered conformation along a bond is thus more stable than the eclipsed and is the conformation that most ethane molecules will have in a given sample.

If a group that is larger than H (usually a methyl), it can be big enough to have steric strain between it and a nearby group on the neighboring C even if they are not eclipsing. This type of steric strain is called a **Gauche interaction**:

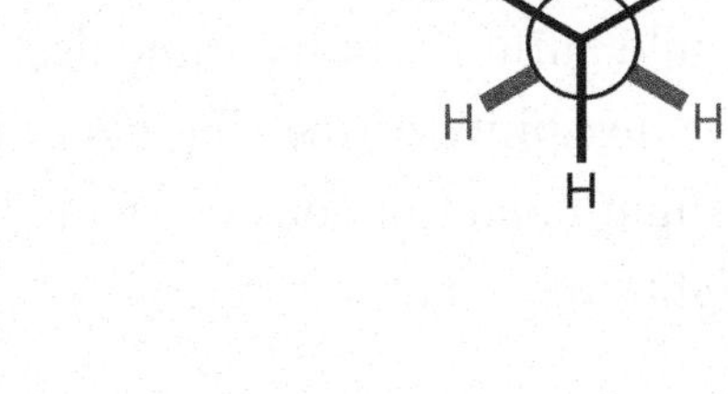

Gauche interaction

Lesson 15. Conformational Analysis II: Cycloalkanes and the Chair Conformation of Cyclohexane

Lesson 15.1. Strain in Cycloalkanes

Cycloalkanes can be subject to constrained geometries that lead to deviations from ideal bond angles. For example, the C atoms in cyclopropane, with equal C–C bond angles, will form an equilateral triangle, with angles of 60°. This angle differs from the ideal angle of 109.5° for an sp^3-hybridized C atom. The 'ideal' angle is the angle that gives the strongest bond, thus holding the molecule together most tightly (i.e., making it most stable). Deviating from the ideal angle thus leads to weaker bonds and less stable molecules. The decreased stability that results from the bonds being strained from the ideal angle is called **angle strain**. The angles that would be present in the planar forms of some cycloalkanes are provided here:

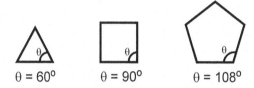

$$\theta = 60° \qquad \theta = 90° \qquad \theta = 108°$$

We saw in the previous Lesson that certain conformations of molecules, specifically those in which there are eclipsing interactions, are less stable than are staggered conformations. When atoms are constrained in a ring, there is not free rotation about a C–C bond, so it is sometimes impossible to alleviate eclipsing interactions in the way we did with ethane and butane. For this reason, there can also be some steric strain induced in cycloalkanes. The sum of all steric and angle strains together is often referred to as **ring strain**.

An example of steric stain induced by molecular geometry is in cyclopropane. Cyclopropane is locked in a conformation in which every H is eclipsing two others (the Newman Projection below shows the perspective down one C–C bond in the right-hand image):

If cyclobutane were planar, it would suffer from the same destabilizing forces (left image, below). As ring size increases, however, the flexibility of the ring also increases. This flexibility allows the cyclobutane to "pucker" a bit (i.e., slight bond rotations occur) in response to eclipsing interactions to alleviate some of the steric strain (right image, below):

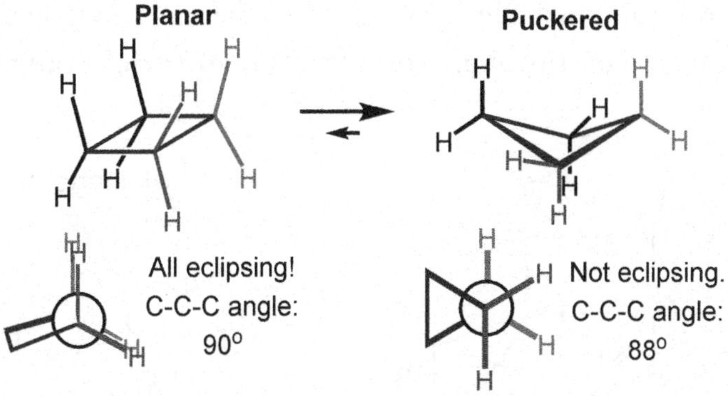

Planar / Puckered

All eclipsing!
C-C-C angle:
90°

Not eclipsing.
C-C-C angle:
88°

Similarly, cyclopentane distorts from a planar conformation to an envelope conformation, which is the actual form found in solution, via what is known as an **envelope distortion** (right image):

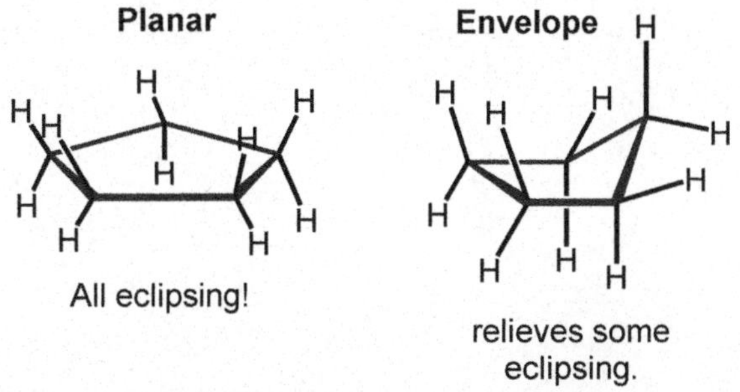

Planar / Envelope

All eclipsing!

relieves some
eclipsing.

The relative strain per CH_2 unit in cycloalkanes trends in the order cyclopentane < cyclobutane < cyclopropane because flexibility increases as the ring size increases (which allows greater C–C bond rotation), thus allowing greater ability to alleviate steric strain. This effect becomes dramatic in the case of cyclohexane, as detailed in the next section.

Lesson 15.2. The Chair Conformation of Cyclohexane

In cyclohexane, the flexibility of the ring allows two envelope distortions to occur (one up and one down) to produce what is called the **chair conformation** of cyclohexane:

fold flap up

'envelope-like'
distortion 1

fold flap
down

'envelope-like'
distortion 2

**Chair conformation
of cyclohexane**

67

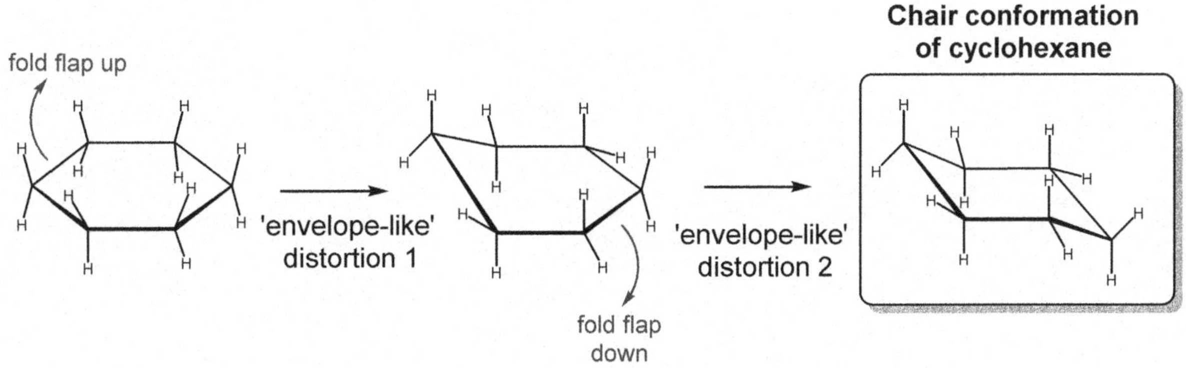

In the chair conformation, all bond angles are ideal, so there is no angle strain. Additionally, there are no eclipsing steric strains. Thus, **the chair conformation of cyclohexane thus has ~0 ring strain:**

All staggered.

Lesson 16. Stereochemistry I: Chirality and Optical Activity

Lesson 16.1 Chirality and Configurational Isomerism

We saw in Lesson 12 that isomers that have the same constitution but cannot be interconverted without bond breakage are called configurational (or geometrical) isomers. The configurational isomers that we discussed were the *cis-* and *trans-*isomers of cycloalkanes. We will now discuss another type of configurational isomerism. A molecule that has "handedness" is said to be chiral. To identify what molecules have "handedness", it might be helpful to think about what makes a right hand and a left hand different. After all, both hands have four fingers and a thumb attached to a palm. No matter how we rotate a right hand, however, it is not superimposable (not identical) with a left hand. They are mirror images of each other. If we think of a hand has being four surfaces, it may help. A hand has 1) a palm, 2) a backhand, 3) a thumb side and 4) a little finger side. If the hand was the same on front and back or if the hand was symmetric on the front (like a cartoon hand with three fingers), then we could superimpose them. However, because all four 'sides' are different, the hands have what we have come to call "handedness". For molecules, we use the term "**chirality**" in place of "handedness". A molecule that possesses chirality is said to be **chiral**. A tetrahedral C atom has four substituents much as a hand has four "sides". If all four substituents coming off the tetrahedral C are different, we say the C atom is a **chiral center** (alternatively known as a **chirality center**, **stereocenter**, or **stereogenic center**, in various books). If any two substituents coming off of the C are identical, the C is not a chiral center. We sometimes indicate stereocenters in a molecule with an asterisk (*).

Example 16.1

Which carbon atoms in these structures are stereocenters?

Solution 16.1

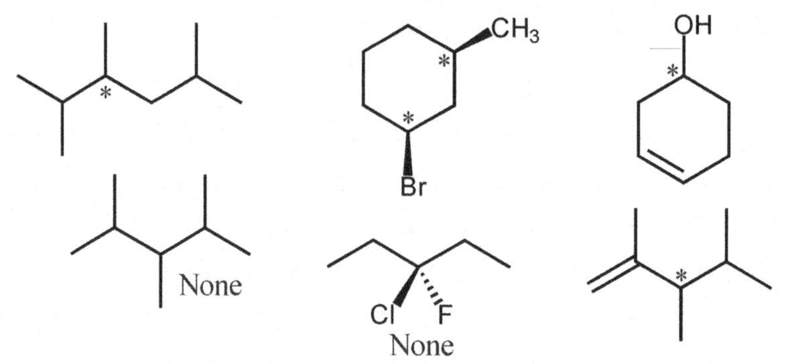

(molecular structures shown: labeled "None", with Br and CH₃, with OH, with Cl and F labeled "None")

Lesson 16.2. Properties of Stereoisomers

A chiral molecule and its mirror image share many of the same chemical properties: the two will have the same stability, solubility, boiling point, and melting point, for example. How, then, do we tell the two apart, and why does chirality even matter? One property of chiral molecules that differs between the two mirror images is the direction in which they rotate plane polarized light. For this reason, a chiral molecule is also said to be **optically active**. An instrument called a polarimeter is used to measure the angle by which plane polarized light is rotated by a given sample (units in °, positive values for clockwise rotations, negative values for counter-clockwise rotations). A simplified drawing of a polarimeter is provided here:

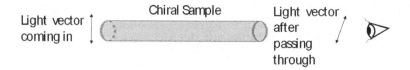

A chiral molecule that causes clockwise rotation of the plane polarized light vector is called a **dextrorotatory** molecule, and a "(+)-" symbol is placed in front of that molecule's name. If the mirror image of this dextrorotatory isomer is measured at the same concentration, we would instead observe a counterclockwise rotation of the light vector by the same angle. The mirror image of dextrorotatory is **levorotatory**, and a "(−)-" symbol is placed in front of a molecule's name to indicate it is a levorotatory molecule. A sample of **achiral** (not chiral) molecules will not cause the plane polarized light vector to rotate at all, and such samples are said to be **optically inactive**. Because the dextrorotatory and levorotatory isomers rotate the vector by the same angle but with opposite signs (i.e., in opposite directions), a sample containing dextrorotatory and levorotatory isomers in equal amounts will exhibit no optical activity (the + and − cancel out). A 1:1 mix of dextrorotatory and levorotatory isomers is termed a **racemic mixture** (or **racemate**) to distinguish it from samples of achiral molecules.

Lesson 17.1: The Cahn-Ingold-Prelog Rules

In the previous Lesson, we observed that the dextrorotatory and levorotatory isomers of a molecule rotate plane polarized light in different directions. The direction of light rotation, however, does not correlate with molecular structure in a predictable way. For this reason, scientists have come up with a system to identify chiral molecules that does not involve rotation of plane polarized light. The naming system is based on the relative spatial orientation of different substituents in order of priority about the chiral center. To name chiral molecules, then, we must first learn the convention for assigning priority to the substituents coming off a stereogenic atom, which is called the **Cahn-Ingold-Prelog** (CIP) rules (See Lesson 12 for a review).

Lesson 17.2. Assigning R- and S- Labels of Configuration

When we label hands, we use the terms "right" and "left" as indicators of the handedness. When naming molecules, we use the Latin terms for right (*rectus*) and left (*sinister*) to indicate molecular handedness. We place a label of configuration in front of the molecule's name (i.e., a stereochemical prefix) so the reader knows to which isomer the name refers: "(*R*)-" for "rectus" and "(*S*)-" for "sinister".

To determine whether an isomer is (*R*)- or (*S*)-, we follow these steps:
1) Assign priorities to the four groups (using CIP rules)
2) Point the **lowest** priority group (4th place) away from you
3) Totally ignore the 4th priority group now and determine the direction of procession from **1→2→3** priorities. If the procession is clockwise, that chiral center has an "(*R*)-" configuration. If the procession is in the counterclockwise direction, that chiral center has an "(*S*)-" configuration.

Example 17.1

Assign a configurational label to this molecule:

Solution 17.1

First assign priorities for each of the substituents:

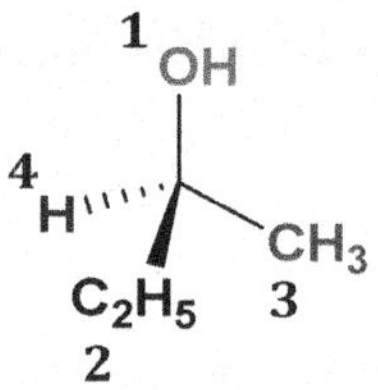

Here, the 4th place substituent is already pointing away from you, so simply draw an arc from 1 to 2 to 3, ignoring the 4th place substituent:

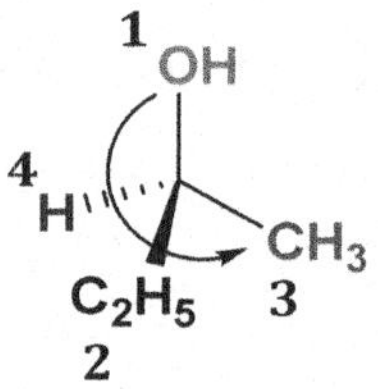

Here, the arc goes in a counterclockwise direction, so this molecule is in the (*S*)- configuration. Its full name would be written as (*S*)-2-butanol.

If the lowest priority substituent is not initially pointing back, you will need to rotate the molecule to point the 4th substituent back. You can do this by building a model and physically turning it or, as you become more proficient with these problems, mentally picturing the molecule and rotating it. One alternative strategy to consider is, if the lowest priority atom is pointing towards you (opposite of what you want), you can instead count 3→2→1 (opposite of normal counting), and you will still attain the correct configuration. If the lowest priority substituent is in the plane of the page, however, this alternate strategy cannot be used, and you will have to reorient the molecule. In the next Lesson, we will learn a convenient representation of chiral molecules drawn such that all substituents are pointed either towards or away from the viewer. Helpful video examples for manipulating 3D structures to determine configuration are provided on the Proton Guru YouTube channel (search "Proton Guru Channel" on youtube.com to find the channel).

Lesson 18. Stereochemistry III: Fischer Projections

Our study of stereochemistry and chirality has made the importance of three-dimensional shapes of molecules evident. The **Fischer Projection** is another way to represent the three-dimensional structure of chiral molecules. In this convention, each stereogenic atom is placed at the intersection of orthogonal lines. The groups on the horizontal lines are understood to represent substituents pointing towards the viewer from the chiral center, with groups on the vertical lines representing substituents pointing away from the viewer from the chiral center.

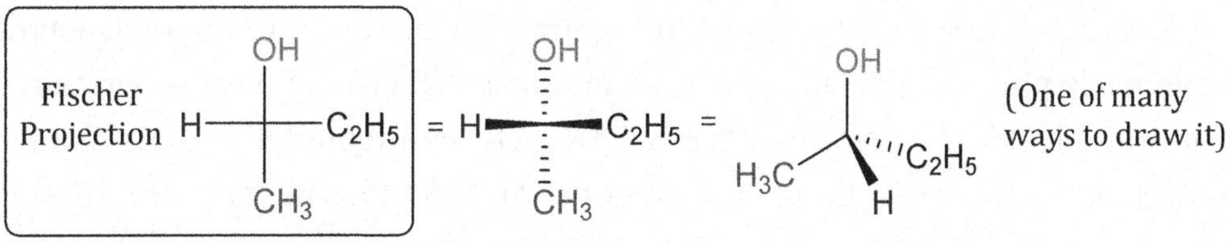

Example 18.1

Assign the configuration of the chiral center in the Fischer Projection shown below:

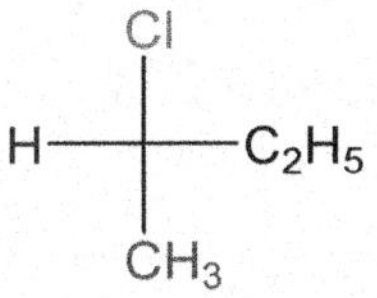

Solution 18.1

First, prioritize the substituents:

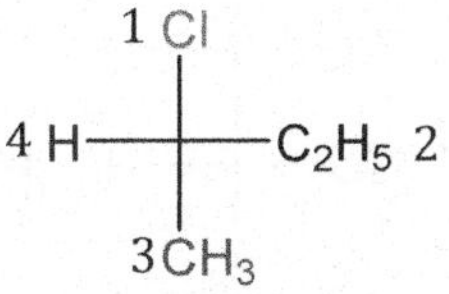

We see that the 4th place substituent (H) is towards us in the Fischer Projection convention. This is a "backwards" molecule, so we count 3→2→1, instead of 1→2→3 (see Lesson 12):

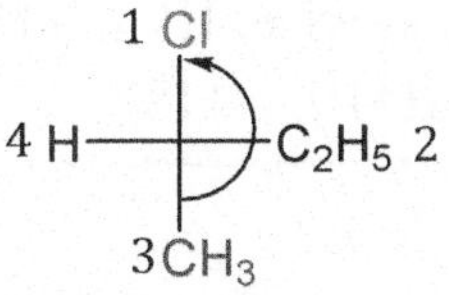

We see a counterclockwise precession from 3→2→1, corresponding to the (*S*)-configuration.

Lesson 19 Stereochemistry IV: Enantiomers, Diastereomers and Meso Compounds

Lesson 19.1. Enantiomers

A molecule may have only one chiral center or it may have many stereocenters. Each stereocenter in a molecule may be in the (R)- or (S)- configuration, so a molecule with "n" stereocenters can have up to 2^n stereoisomers. As an example, a molecule with two stereocenters at carbons 1 and 2 can have up to 2^2, or 4, stereoisomers: the "(1R,2R)-" isomer, the "(1S,2S)-" isomer, the "(1R,2S)-" isomer, and the "(1S,2R)-" isomer. Some vocabulary is needed to delineate the relationship between these isomers. Two chiral molecules that are non-superimposable mirror images are referred to as **enantiomers** and form an **enantiomeric pair**. A molecule and its enantiomer will have opposite configurations at *every* stereocenter in the molecule. If a molecule has a "(1R,2R)-" configuration, its enantiomer will have a "(1S,2S)-" configuration. Likewise, the enantiomer of a "(1R,2S)-" isomer is a "(1S,2R)-" isomer.

Lesson 19.2. Diastereomers

Stereoisomers that are not mirror images are known as **diastereomers**. In contrast to enantiomers, two diastereomers do not necessarily have the same stability, solubility, melting point, boiling point, etc. Consequently, it is possible to separate diastereomers by distillation, recrystallization, etc., much easier than for enantiomers. For example, "(1R,2R)-" isomer will be a diastereomer of both the "(1R,2S)-" and the "(1S,2R)-" isomers.

Example 19.1

Provide the stereochemical relationships between the compounds shown below:

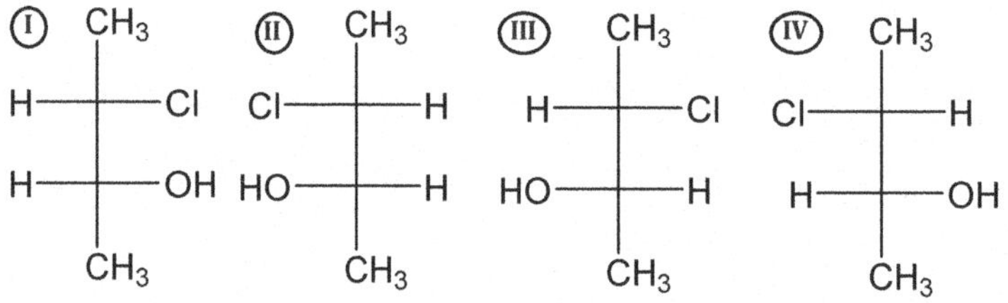

Solution 19.1

We can see that I and II are mirror images of one another and are non-superimposable, so I and II form an enantiomeric pair. Likewise, III and IV are enantiomers. All other pairs we compare are diastereomers (I+III, I+IV, II+III, II+IV).

Lesson 19.3 Meso Compounds

It is possible for a molecule to contain stereogenic *atoms* in its structure, but that the *molecule as a whole* is achiral. This condition occurs when one half of a molecule is the mirror image of the other half (i.e., has a plane of symmetry). Remember, a symmetric object cannot be chiral. Molecules that have stereogenic atoms but which are achiral molecules are called **meso compounds**.

74

Example 19.2

Which of the following are meso compounds?

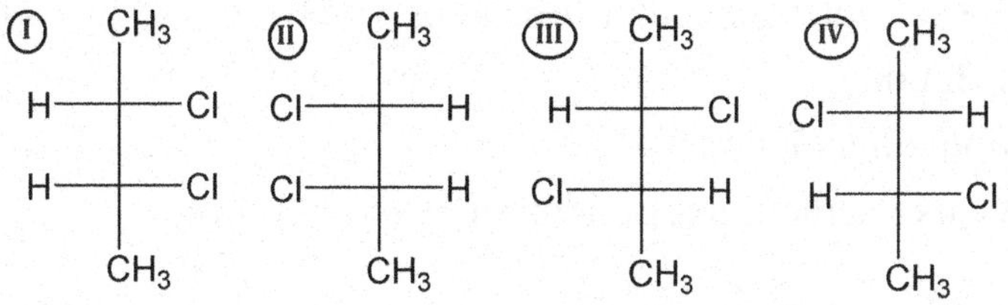

(I)
```
        CH3
        |
  H ----+---- Cl
        |
  H ----+---- Cl
        |
        CH3
```

(II)
```
        CH3
        |
 Cl ----+---- H
        |
 Cl ----+---- H
        |
        CH3
```

(III)
```
        CH3
        |
  H ----+---- Cl
        |
 Cl ----+---- H
        |
        CH3
```

(IV)
```
        CH3
        |
 Cl ----+---- H
        |
  H ----+---- Cl
        |
        CH3
```

Solution 19.2

Compounds I and II each possess a plane of symmetry (represented by the dashed lines) and are thus achiral, despite the fact that carbons 2 and 3 are stereocenters. Compounds I and II are therefore meso compounds. In fact, they are identical.

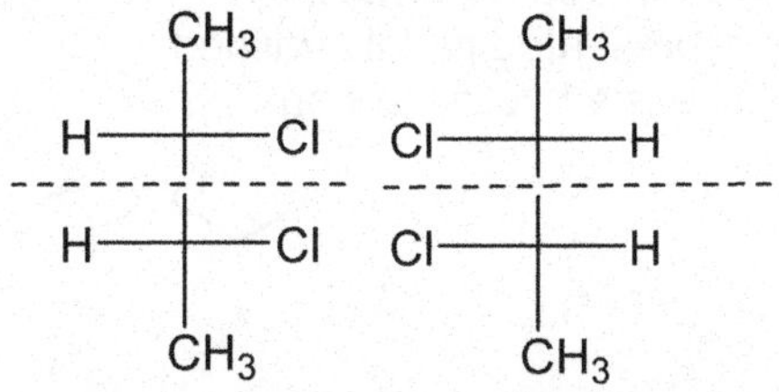

```
        CH3                    CH3
        |                      |
  H ----+---- Cl        Cl ----+---- H
- - - - - - - - - - - - - - - - - - - - -
  H ----+---- Cl        Cl ----+---- H
        |                      |
        CH3                    CH3
```

Lesson 20.1. Intermolecular Forces

There are three types of intermolecular forces on which we will focus in this book:

    i.   Hydrogen Bonding

    ii.   Dipole–Dipole interactions

   iii. van der Waals interactions (London dispersion forces).

The **strongest** of these intermolecular forces is **hydrogen bonding** (sometimes abbreviated "H-bonding"). Hydrogen bonding is a specific type of Coulombic attractive force between the partial negative charge of one polar bond and the partial positive charge on an H atom (induced by the H atom being in a very polar bond):

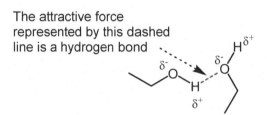

The attractive force
represented by this dashed
line is a hydrogen bond

The greater the charges on H and X in the molecule, the greater the Coulombic attraction will be, so the two molecules will be more strongly attracted to one another. The elements with the three highest electronegativity values are F > O > N. Consequently, the most polar H–element bonds occur in molecules in which hydrogen is bonded to one of these elements. Molecules capable of the strongest H-bonds all feature H–F, H–O and H–N bonds. For the purposes of organic chemistry courses, you will typically only see compounds with H–O and H–N bonds engaging in hydrogen bonding.

The **second strongest** of the intermolecular forces we will cover are **dipole–dipole interactions**. Dipole–dipole interactions are simply the attractive forces between a partial positive charge on an atom in one molecule and the partial negative charge on an atom in another molecule. The greater the partial charges, the stronger the Coulombic attraction between the two:

The attractive force represented by the dashed line
is a **dipole-dipole interaction**

The weakest of the intermolecular forces we will cover are van der Waals interactions, sometimes called London Dispersion Forces. The van der Waals interactions arise from temporary dipoles that form in molecules that lack polar bonds. When two such molecules come in close proximity, the electron clouds on the atoms in one molecule repel the electron clouds on the atoms in the other molecule. This will, very briefly, cause a very weak 'induced dipole' on the molecules whose electrons were repelled. These forces, however, are far weaker than either dipole–dipole interactions or H-bonds.

Lesson 20.2. Using Intermolecular Forces to assess Relative Melting and Boiling Points

In a solid, it is the intermolecular forces that hold molecules close together with enough force to allow the solid to maintain its shape. A solid will only melt into liquid form when the intermolecular forces are disrupted enough to allow the molecules to flow past one another. To disrupt intermolecular forces, one needs to add energy, generally by heating the sample to its melting point. *The stronger the intermolecular forces, the more heat (higher melting point) is needed* to melt the sample. This observation allows us to predict the relative melting points of several samples simply by comparing their structures.

The boiling point of a liquid can be estimated in a similar way. In a liquid, the molecules remain close to one another even though the molecules can flow past one another. When enough energy is added to boil the liquid, the intermolecular forces are completely overcome, allowing the molecules to break free into the gaseous state, whereupon the molecules too far apart to experience intermolecular forces between one other. *The stronger the intermolecular forces, the more heat (higher boiling point) is needed* to boil the sample.

Example 20.1

List the strongest intermolecular force that is present between molecules in a sample of each of the following and rank the compounds 1–4 in terms of boiling point, 1 being highest.

Solution 20.1

The strongest intermolecular force in each is:

I: H-bonding (two sites)

II: dipole-dipole

III: van der Waals interactions (London dispersion forces)

IV: H-bonding (one site)

The boiling point increases as the strength of the force increases, and compound I has *two* H-bonding units whereas compound IV only has one. So, the order of boiling point is **I > IV > II > III**, where compound **I** has the highest boiling point

---

77

Lesson 20.3. Effect of Molecular Weight and Branching on Boiling and Melting Point)

We have seen that the type of intermolecular forces in a sample has a dramatic effect on the boiling and melting points. What if we have to compare two samples that have the same type of intermolecular forces? For example, hydrocarbons are nonpolar, so they only have dispersion forces. To compare the b.p. or m.p. of two hydrocarbons, we need additional rules. First, **the greater the molecular weight** (# of carbons) of the molecule, **the higher the boiling point or melting point**:

Molecule	B.p. (°C)
C_4H_{10}	−0.5
C_5H_{12}	36.1
C_6H_{14}	68.7
C_7H_{16}	98.4
C_8H_{18}	125.7

Second, **the larger the surface area the higher the boiling point and melting point**. This means that linear alkanes have higher b.p./m/p. compared to branched alkanes of the same mass:

C_6H_{12}		
b.p =36.1°C	b.p =27.9°C	b.p =9.5°C

Lesson 20.4. Predicting Solubility and Classifying Solvents

In addition to the boiling point and melting point, the type of intermolecular forces of which a molecule is capable also influences solubility. The old chemistry saying "Like Dissolves Like" is a good rule of thumb for predicting solubility: polar molecules dissolve in polar solvents and nonpolar molecules dissolve in nonpolar solvents. The better the match of intermolecular force type, the higher the solubility. Solvents can first be grouped into polar solvents (these can H-bond or have dipole-dipole interactions) and nonpolar solvents (these only have dispersion forces):

Nonpolar Molecules	Polar Molecules
Alkanes	Alcohols
Alkenes	Ethers
Alkynes	Amines
Aromatic	Carboxylic acids

Polar solvents can be further categorized into two types: **Protic** (these can be H-bond donors) and **Aprotic** (these cannot be H-bond donors but have dipole-dipole interactions). The polarity of solvents is quantified as the dielectric constant (ε). Dielectric constant (ε) is a measure of the ability of material to moderate the force of attraction between oppositely charged particles. Some examples of dielectric constants for a few common lab solvents are provided here:

Polar Solvent	Abbreviation	Structural Formula	Protic or Aprotic	Dielectric constant (ε)
Water	H_2O	H–O–H	Protic	80
Dimethyl Sulfoxide	DMSO	$H_3C-\overset{\overset{O}{\|\|}}{S}-CH_3$	Aprotic	49
Acetonitrile	ACN	$H_3C-C{\equiv}N$	Aprotic	37
N,N-dimethylformamide	DMF	$H_3C-\underset{\underset{CH_3}{\|}}{N}-\overset{\overset{O}{\|\|}}{C}-H$	Aprotic	37
Methanol	MeOH	H_3C-O-H	Protic	33
Acetic Acid	AcOH	$H_3C-\overset{\overset{O}{\|\|}}{C}-OH$	Protic	6
Diethylether	Et_2O	$H_3CH_2C-O-CH_2CH_3$	Aprotic	4

Increase in polarity

Some other good rules of thumb for comparing relative solubilities are:

1) For molecules with similar MW, the **higher** the polarity (dipole-dipole / H-bonding) the **higher** the solubility in **polar** solvents:

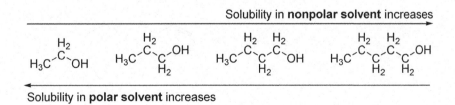

2) As the hydrocarbon chain on polar molecule increases, the solubility in polar solvents **decreases** and solubility in nonpolar solvent **increases**:

Solubility in **nonpolar solvent** increases

Solubility in **polar solvent** increases

Lesson 21. Boiling and Melting Point Trends in Alkenes

We saw in the previous lesson that the types of intermolecular forces in a sample can influence the melting and boiling points and that the shape (like linear versus branched) also has an effect. Unsaturated hydrocarbons, like their saturated counterparts, lack heteroatoms and thus do not contain bonds with significant polarity. As a result, the intermolecular forces (IMFs) in alkenes are dominated by London dispersion forces (LDFs).

In the last lesson, we saw that increasing the size or linearity of an alkane led to increased IMFs/higher boiling points, and this trend is also observed with alkenes. For example, 1-hexene has a higher boiling point than 1-pentene (30 °C vs. 63 °C, respectively). Similarly, the more closely that alkene molecules can pack together, the stronger the IMFs will be. For alkenes, more rigid alkenes can pack more effectively than can flexible alkenes, just a box of rigid pencils can pack well, whereas a box of flexible pieces of rope forms a less well-packed assembly. The added conformational flexibility about sp^3 hybridized C3 in 1,4-pentadiene is thus the reason that its boiling point (26 °C) is lower than that of *trans*-1,3-pentadiene (42 °C). The stereochemistry of the C=C bond also influences the IMFs and properties of alkenes. For example, molecules of more linear *trans*-2-butene can pack together more closely than molecules of U-shaped *cis*-2-butene, which can be observed by the higher melting point of the *trans*-isomer:

	1-pentene	bp = 30 °C
	1-hexene	bp = 63 °C
	1,4-pentadiene	bp = 26 °C
	trans-1,3-pentadiene	bp = 42 °C
	trans-2-butene	mp = −105 °C
	cis-2-butene	mp = −139 °C

Example 21.1

Which of the following alkenes would have the higher melting point?

vs

<u>Solution 21.1</u>

Both of these molecules are heptenes, so it cannot be molecular weight that causes them to have different melting points. However, we can see that the molecule on the left (2-methyl-1-hexene) has one branch, whereas the molecule on the right (1-heptene) has no branches. As a result, 1-heptene is able to pack together more densely than 2-methyl-1-hexene. Because the molecules of 1-heptene are able to pack more densely, the IMFs between the molecules are stronger. Because the IMFs between the 1-hexene molecules are stronger, it will require more energy to pull the molecules apart, which will be manifest in the form of a higher melting point.

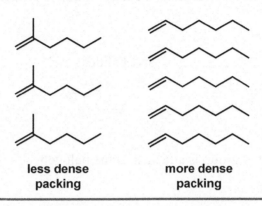

less dense
packing   **more dense**
                   packing

Lesson 22.1. Types of Alkyl Halides

Alkyl halides undergo two important types of reactions: nucleophilic substitution (S_N) and elimination (E). The basic principles and terms used to describe these reactions are used to understand other, more complicated reactions exhibited by other classes of compounds as well, so this is an important topic. If we are going to make decisions or predictions about alkyl halides we need to familiarize ourselves with the basic types and classifications of them.

Types

alkyl halides - RX   aryl halides - ArX   vinyl halides

CH_3Br CH_3CH_2Cl

—Br

CH_2=CHCl

Classification

methyl halide - a methyl group is attached to the halogen.

CH_3Br

primary (1°) alkyl halide - the carbon with the halogen is attached to one alkyl group.

CH_3CH_2Cl

look at this carbon

—Br

—Br

look at this carbon
R-CH$_2$-Br

secondary (2°) alkyl halide - the carbon carrying the halogen is attached to two alkyl groups.

Br

look at this carbon

Cl

R_2-CH-X

tertiary (3°) alkyl halide - the carbon attached to the halogen is joined to three alkyl groups.

Br   Br R_3C-Br

In biochemistry, a **substrate** is the molecule upon which an enzyme reacts or the surface where an organism grows. In organic chemistry, the *substrate* is a reactant – the starting material – that is altered with a reagent and converted to the product of interest. The above types of alkyl halides are usually the substrates in the S_N and E type reactions. Some other terms used in these reactions need to be described.

A typical nucleophilic substitution (S_N) reaction displaces one group and replaces it with another group.

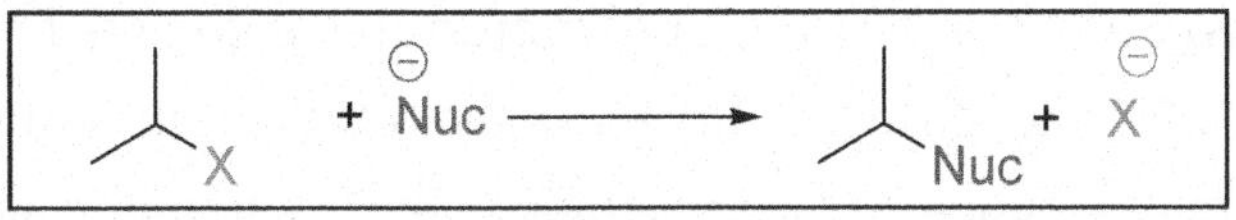

The terms we will be using throughout the semester are shown below.

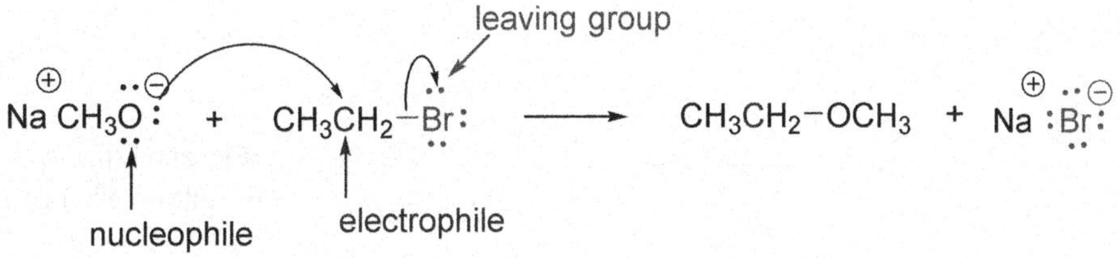

The *nucleophile* (attracted to positive sites) is the species that attacks the alkyl halide (*substrate*):

- most nucleophiles are anions

- some nucleophiles are neutral

- an electron pair is required

Some typical nucleophiles are shown:

- sulfur: HS^-, RS^-
- oxygen: HO^-, RO^-, RCO_2^-
- carbon: CN^-, $HC \equiv C:^-$

halides: Br^-, I^-, Cl^-, F^- (although halides are typically leaving groups)

- nitrogen: N_3^-

The leaving group is any group that can be displaced from a carbon atom. We'll discuss these again later and we will analyze them similar to the way we analyzed conjugate bases in Lessons 6-8.

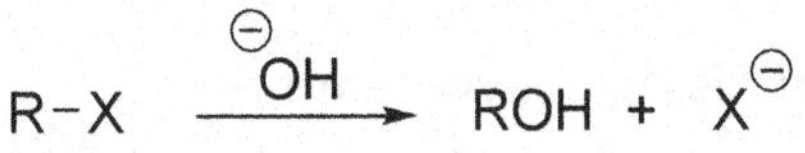

Leaving group (X)	relative rate
I	30,000
Br	10,000
Cl	200
F	1

Earlier, we compared acid strength by looking at the stability of the conjugate base. In comparing HI, HBr, HCl, and HF we said that HI was the strongest acid because the negative charge on the conjugate base, I^-, was more diffuse than the others due to size. When connected to a hydrogen atom it's called a conjugate base, when connected to a carbon atom, it's called the leaving group. But the reason why I^- is a better leaving group is the same: a more diffuse negative charge because I^- is larger.

S$_N$2 stands for *substitution nucleophilic bimolecular* and it refers to a **concerted** (meaning one-step) process. It "*substitutes*' the halide (leaving group) with the *nucleophile*. The bimolecular term is a kinetics reference meaning that the rate depends on two reactants – the alkyl halide and the nucleophile. The rate law for this reaction would be this: rate = k[RX][Nuc$^-$]. The "2" in S$_N$2 also means that reaction depends on or only occurs when "2" species collide. In general chemistry terms, the rate-determining step of the reaction is bimolecular.

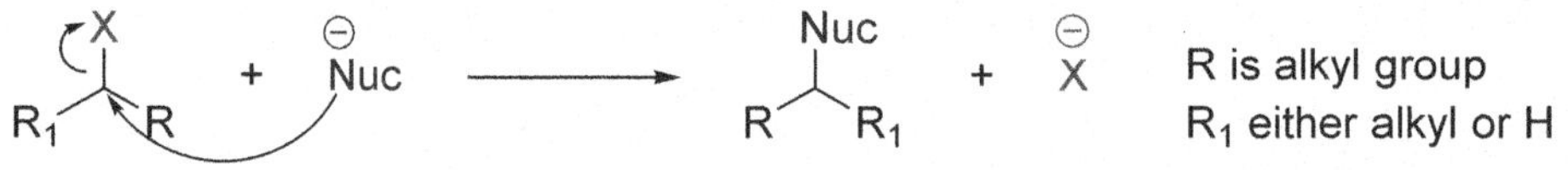

R is alkyl group
R$_1$ either alkyl or H

Above, we hinted at the impact of the leaving group on a reaction's rate, or viability. The substrate also influences the rate of a S$_N$2 reaction.

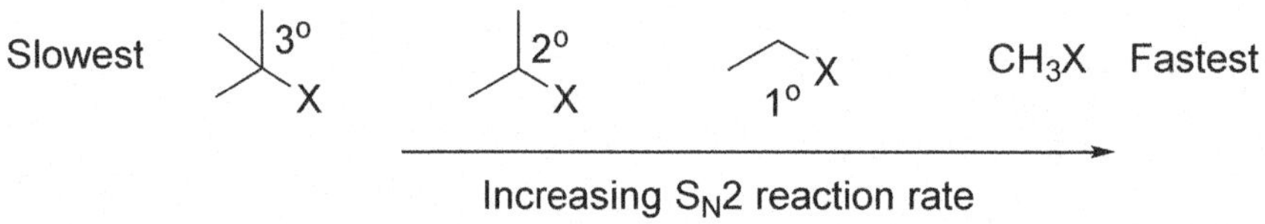

Increasing S$_N$2 reaction rate

Notice that we didn't extend the arrow back to the tertiary alkyl halides, that's because it is so crowded around a tertiary carbon atom that the nucleophile is blocked from gaining access. Tertiary alkyl halides do undergo substitution reactions; they just don't go through the S$_N$2 pathway. The reason for the observed rate trend is called *steric hindrance*. Steric hindrance refers to the bulkiness or crowdedness around the alkyl halide carbon that is being attacked (the electrophile). The concept of steric hindrance can also be applied to nucleophiles: the bulkier the nucleophile, the slower the reaction.

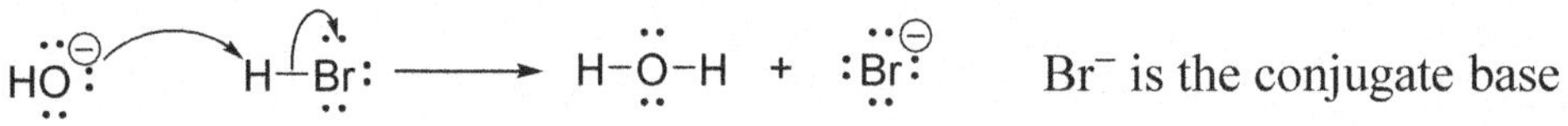

Br$^-$ is the conjugate base

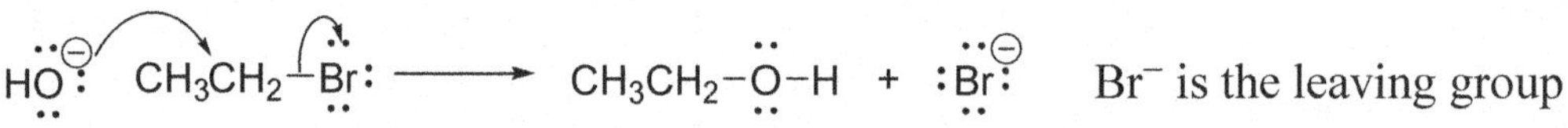

Br$^-$ is the leaving group

In general chemistry, particularly kinetics, you learned that the success of a reaction depends on two things – 1) a sufficient amount of energy in the collision (some refer to this as frequency factor), and (2) the proper orientation. These considerations were hidden in the pre-exponential factor, A, of the Arrhenius equation, k = Ae$^{-Ea/RT}$. We're not here to revisit the math, just to show that organic chemistry approaches this from a somewhat visual, conceptual perspective. A typical S$_N$2 reaction and its reaction coordinate diagram are shown below.

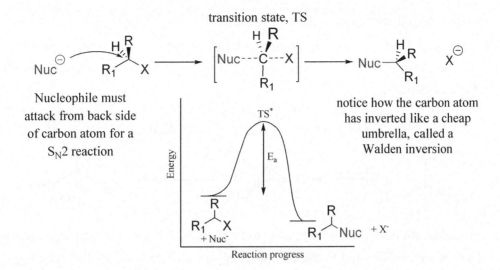

No math here, just two species colliding with the nucleophile attacking from the side opposite the side to which the leaving group is attached, a feature known as *backside attack*. This properly-oriented collision causes the leaving group (a halide ion here) to leave as the carbon atom undergoes an *inversion of configuration*. This inversion is often called *Walden inversion*. We also assume here that the collision has enough energy to get over the activation energy barrier in the process.

To summarize the S_N2 reaction:
1. The rate depends on the concentration of the nucleophile and the substrate.

   rate = k[Nuc][R−X]
2. The S_N2 reaction occurs with an inversion of configuration (stereochemistry).
3. The S_N2 reaction is fastest when the substrate is methyl or primary due to the lack of steric hindrance.

Lesson 24. The S$_N$1 Mechanism and the Carbocation

We are now going to learn our first reaction mechanism that involves more than one step: the S$_N$1 reaction. The S$_N$1 reaction involves: 1) heterolysis to form a carbocation, then 2) coordination of a nucleophile to the carbocation:

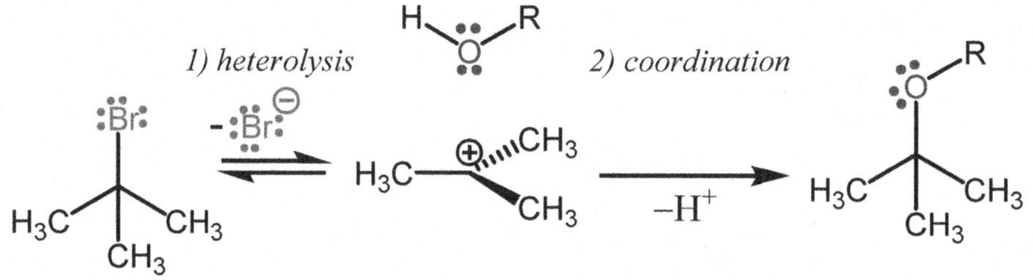

Note that the carbocation to which the nucleophile coordinates is very reactive, so that even a weak nucleophile – like the neutral alcohol in the example above – will suffice for this reaction.

Here are some examples of net S$_N$1 reactions, without showing the mechanistic steps:

General:
$$R\text{—}LG \;+\; H\text{—}Nu \longrightarrow R\text{—}Nu \;+\; H^+ + LG^-$$
$$R\text{—}LG \;+\; Nu^- \longrightarrow R\text{—}Nu \;+\; LG^-$$

Br$^-$ as LG
H-OH as H-Nu

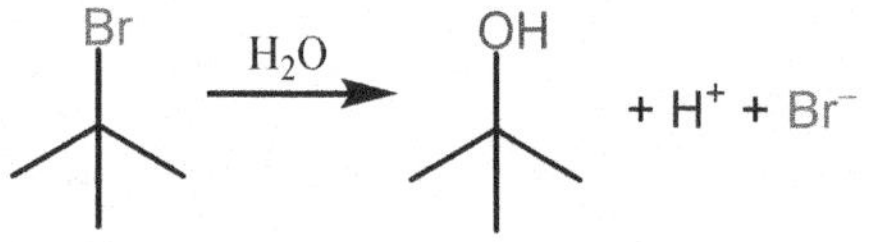

+ H$^+$ + Br$^-$

note that H$^+$ comes off of the H$_2$O after H$_2$O adds so that the O will be neutral

H$_2$O as LG
Br$^-$ as Nu$^-$

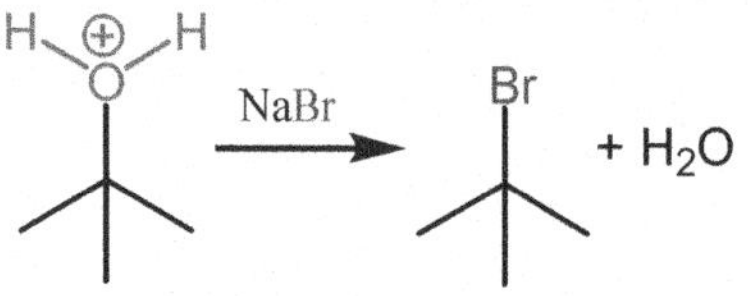

+ H$_2$O

note that NaBr dissociates in solution to form Na$^+$ and Br$^-$ and that the LG in this case is a neutral molecule

The net result is that the *leaving group* (LG) – the group removed from C in the heterolysis step – is substituted by the nucleophile. The "S" in S$_N$1 stands for substitution and the subscript "N" stands for nucleophilic. The "1" indicates that the kinetics of the reaction are *unimolecular*: the reaction rate only depends on the concentration of *one* species. In this case, we look at the mechanism and see that the carbocation is the least stable species formed on the way to the product, so this step is *rate-limiting*. A qualitative reaction coordinate diagram for a thermodynamically-favorable S$_N$1 reaction might look like this:

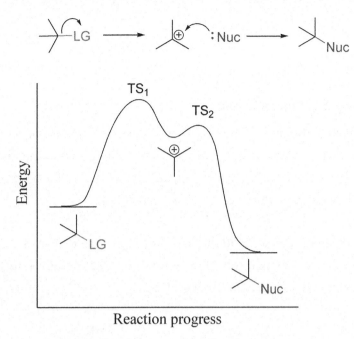

Where **(CH₃)₃C−Br** is the reactant, **(CH₃)₃C⁺** is the carbocation intermediate. the peaks **TS₁** and **TS₂** are transition states, and **(CH₃)₃C−Nuc** is the product. The first 'hill' represents the activation barrier for heterolysis and the second 'hill' represents the activation barrier for coordination.

The rate of the overall two-step process is governed by the high E_a of the heterolysis step, so the rate is only dependent on the concentration of the starting material (the substrate, R–LG), so

$$\text{Rate} = k[\text{R–LG}]$$

Note that, since the nucleophile does not show up in the rate law expression, its concentration does not influence the rate. If we double the nucleophile concentration, the reaction rate is unchanged. If we double the concentration of the substrate (R–LG), which *is* a term in the rate law, the reaction rate will double.

The rate of the S$_N$1 reaction can also be influenced by the leaving group identity, substrate substitution degree, and reaction solvent. The better the leaving group (the more stable the species displaced from the substrate), the faster the reaction will be because there is a lower activation barrier to leaving group departure.

In the S$_N$1 mechanism, carbocation formation is rate-limiting, and more stable carbocations form faster. So, in terms of S$_N$1 reaction rate: methyl halide < 1° alkyl halide < 2° alkyl halide < 3° alkyl halide (fastest). Indeed, **the carbocations that would form from methyl and 1° alkyl halides would be so unstable that methyl and 1° halides will never undergo S$_N$1 reactions.**

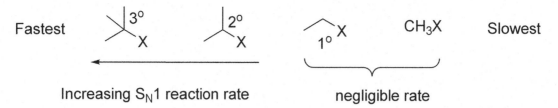

Increasing S$_N$1 reaction rate   negligible rate

So, the substrate effect on the rate of the S$_N$1 reaction is the opposite of the trend for the S$_N$2. The S$_N$2 rate is improved with less steric hindrance, so what's the explanation for S$_N$1? Since the rate-determining step depends on the halide (leaving group) leaving on its own volition, the more stable carbocation is more readily formed, and hence the faster reaction.

When discussing acids and bases we said that alkyl groups are electron-donating groups (EDG) but can only donate electron density inductively. For anions this is a destabilizing effect but inductively donating electron density to a cation is a stabilizing effect.

				arrows note R-group's ability to inductively donate electron density to stabilize the positive charge
methyl	1°	2°	3°	
less stable			**more stable**	

The inductive effect is the easiest way to remind yourself of a carbocation's relative stabilities, but the most current explanation for carbocation stability is hyperconjugation. Since a carbocation only has three electron domains, its hybridization is sp^2, leaving an empty p-orbital (shown below). Because electrons attract positive charge, a σ-bonding electron pair will be attracted to the empty p-orbital on the carbocation, which is a stabilizing interaction. This specific interaction is called hyperconjugation. Due to hyperconjugation, the more non-H substituents on the cationic C, the more stable it is.

General structure of a carbocation

empty p sp^2-hybridized carbon   empty p

σ-bond

------ = hyperconjugation (attractive force)

Let's get back to the S$_N$1 mechanism. Despite S$_N$1 having a "1" in the name, it goes through two steps. We established that it goes through a carbocation intermediate and their relative stabilities and rates. We have not discussed the stereochemistry of the reaction. In the S$_N$2 reaction, done in one step, the incoming nucleophile does a back-side attack and the substrate undergoes a Walden inversion (like a cheap umbrella folding the wrong way in the wind). For the S$_N$1 reaction, look at the general structure of the carbocation. The *p*-orbital has a lobe above and a lobe below the trigonal plane of the alkyl groups; the positive charge is in both lobes of this *p*-orbital. In other words, the nucleophilic attack of step 2 can attach itself to the carbocation from either side.

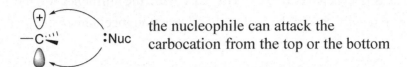 the nucleophile can attack the carbocation from the top or the bottom

So, we say that there is no inversion of configuration for S$_N$1 reactions.

Lesson 25. Factors Determining S_N1 or S_N2 Path

Lesson 25.1. Substrate

In this Lesson, we will discuss four factors that influence whether a reaction will proceed through a S_N1 mechanism or a S_N2 mechanism. We will start with the influence of the substrate. We have established the relative reactivities of the S_N1 and S_N2 reaction mechanisms:

S_N2:

due to steric hindrance

S_N1:

due to carbocation stability

If our alkyl halide is anything but secondary, we can conclude what the most likely mechanistic pathway will be. Those secondary alkyl halides will require consideration from the other factors.

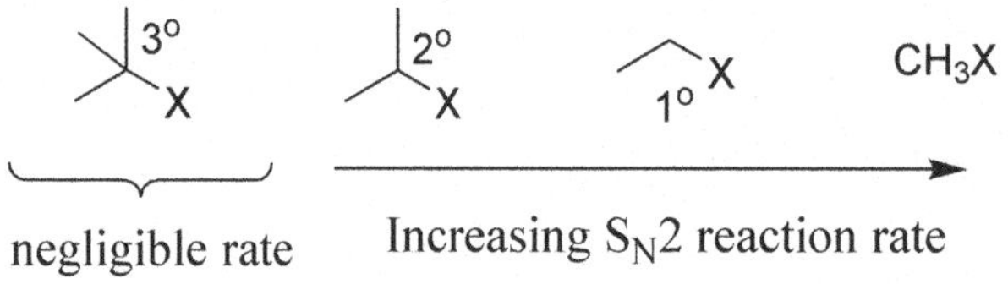

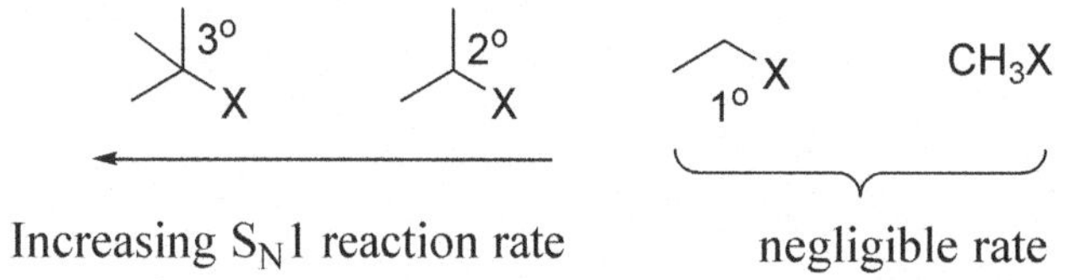

Lesson 25.2. Nucleophile

First, let's correlate some structural features to the relative strengths of nucleophiles:

- A negatively charged anion is a better nucleophile than its neutral conjugate acid. For example, HO^- is a better nucleophile than H_2O; RO^- is better than ROH.
- Larger atoms are better nucleophiles. The iodide ion, I^-, is better than Br^-; $RS^-$is better than RO^-.
- Neutral compounds, like H_2O, ROH, NH_3, and RNH_2, are poor nucleophiles. While they possess the required lone pair of electrons, a full negative charge is better. This lack of charge makes these poor nucleophiles good choices for promoting S_N1 reactions.
- Sterically hindered, bulky anions are poor nucleophiles.

It is easy to confuse nucleophilicity with basicity because bases and nucleophiles both function as electron donors. **The faster a species donates its electron pair to an electrophile, the better a nucleophile it is**. Note the difference in terminology as well: bases are referred to as being 'strong' or 'weak', whereas nucleophiles are classified as 'good' or 'poor'. Let us begin discussing anions that are good nucleophiles. Because nucleophilicity is not tied to stability in the way that basicity is, many more anions are good nucleophiles than are strong bases. What factors might make an anion a poor nucleophile? One factor is electronegativity. Fluorine is the most electronegative element, so it is slow to give up its electrons, making fluoride a poor nucleophile. The steric bulk of a nucleophile also influences its ability to rapidly give electrons to an electrophile. Consider the difference between the nucleophile approach by the small HO^- vs. the bulky $t\text{-}BuO^-$ trying to reach the $\delta+$ of ethyl chloride:

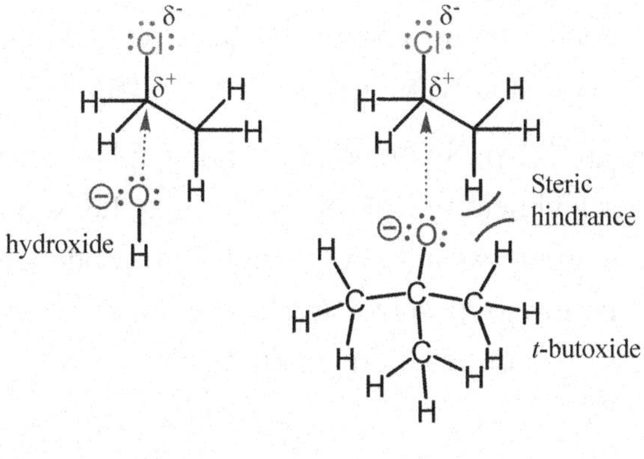

It is not difficult to see that the bulkier $t\text{-}BuO^-$ must overcome a greater a steric repulsive force (leading to a higher energy of activation) than the smaller HO^-. In general, **bulky anions are poor nucleophiles**. "Bulky" in this context means that the sites adjacent to the anionic atom have three or more non-H branches ($t\text{-}BuO^-$ has three methyl branches, for example). To summarize: **non-bulky anions other than fluoride are mostly good nucleophiles**.

Now consider neutral compounds. Most common neutral compounds are significantly more stable than the hydroxide anion (our prototypical strong base), so there are few *strong* neutral bases. Nucleophilicity, however, only requires rapid donation of electrons. Large atoms (3rd row of periodic table or further down, e.g., P and S) with lone pairs can thus serve as good nucleophiles. These atoms are not as electronegative as are the second-row atoms like O and N, so there is a lower activation barrier for pulling electrons from such species.

To summarize our evaluation of nucleophiles:

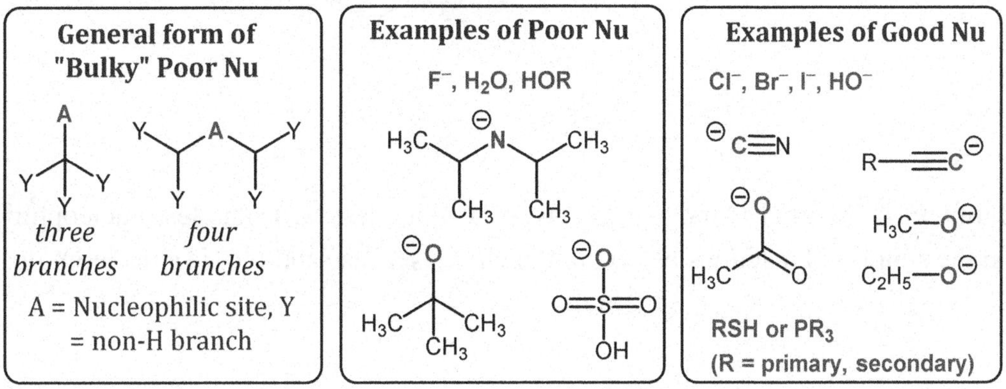

91

Lesson 25.3. Leaving group

We can evaluate leaving group ability in the same way we looked at the conjugate bases upon their departure. The better leaving group is the more stable molecule or ion upon leaving the substrate. As a reminder, weak bases make good leaving groups. What was it that made something a weak base? Having a diffuse, stabilized negative charge or having no charge made for weak bases (see Lesson 7 to review). This same criterion affords a good leaving group. Bad leaving groups are strong bases and have a localized negative charge. RO^-, HO^- and R_2N^- are examples of bad leaving groups.

It's difficult to say whether an S_N1 or S_N2 reaction is favored by considering the leaving group. Very stabilized negative charges and neutral leaving groups may favor S_N1 reactions, but just barely. The main point here is that some groups just do not want to leave, but sometimes we can make a bad leaving group into a good leaving group by some step before it leaves, so we have a lot of leaving groups to choose from when deciding on reaction conditions!

Lesson 25.4. Solvent Effects

Here is the rule you need to know: **polar aprotic solvents favor S_N2**. Let's consider why this might be. Bases and nucleophiles must both donate electron pairs in the course of their usual reactivity, so any factors that hinder electron pair donation will diminish their basicity and nucleophilicity. One way the ability to donate an electron pair is diminished is by strong attraction of solvent molecules for the lone pair or negative charge. The strongest type of interactions that can occur between a neutral solvent molecule and an anion (whether it is a base or a nucleophile) is hydrogen bonding. It is easy to see how the **solvent cage** surrounding an anion could influence its basicity and/or nucleophilicity. Consider A^- solvated by water:

The pull of the $H^{\delta+}$ in the H_2O molecules makes the anion less basic and less nucleophilic, as can be visualized in the simplified reactions below, in which B^- is a base and Nu^- is a nucleophile:

The weakening of basicity/nucleophilicity is most pronounced in H-bonding solvents, which are often referred to as **polar protic solvents**. Common polar protic solvents include water and alcohols.

Polar solvents that cannot engage in H-bonding of bases/nucleophiles are called **polar aprotic solvents**.

Common polar protic solvents: H_2O, ROH, NH_3

Common polar aprotic solvents:

acetone	DMSO	DMF	DME
	dimethyl sufloxide	dimethyl formamide	dimethoxyethane

Although polar aprotic solvents are incapable of H-bonding, they can engage in dipole–ion interactions with the anion. Dipole–ion interactions are weaker than H-bonding to ions, so attenuation of basicity/nucleophilicity is not as pronounced in polar aprotic solvents as it is in polar protic solvents.

Nonpolar solvents (like alkanes) do not engage in strong intermolecular interactions with anions, so basicity and nucleophilicity are highest in these solvents. The general trend is:

nonpolar solvent	<	polar aprotic solvent	<	polar protic solvent
(most basic/nucleophilic)		S_N2 favored		(least basic/nucleophilic)
				S_N1 favored

From a practical standpoint, very polar species that typically dissociate into ions upon dissolution (i.e., ionic salts) are usually insoluble in nonpolar solvents. For this reason, chemists most often do reactions requiring a strong base or a good nucleophile in a polar aprotic solvent to optimize reactivity while maintaining the solubility of their reagents.

Students often struggle recognizing when the leaving group is a neutral molecule because when this happens we need an extra step before the "main" reaction can happen. The bonding electrons depart with the leaving group so a neutral leaving group will have a positive charge before leaving. For alcohols, the hydroxide ion, HO$^-$, never leaves (it is a strong base, after all). We can, however, put a proton on the −OH so that if it leaves it will be water, which is a stable molecule and thus a good leaving group:

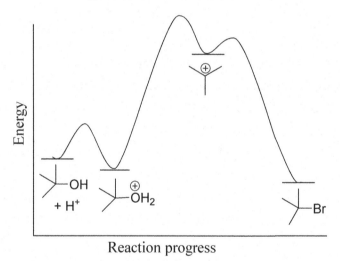

HO$^-$ is a bad leaving group

H_2O is the LG now, and H_2O is a good leaving group... and it does

Look. This Br-atom has a negative charge, perhaps it might act as a nucleophile and coordinate with the carbocation next.

This is a three-step mechanism now, so an energy diagram would have three hills. The first step protonates the −OH, then the normal S$_N$1 mechanism occurs.

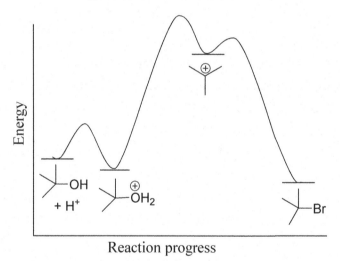

Reaction progress

Lesson 25.6. Recapping Factors Influencing Rate of S_N1 and S_N2 Reactions

Let's summarize the factors and their influence in a table.

	S_N1 rate	Comment	S_N2 rate	Comment
Substrate	3° > 2°	S_N1 needs to form a carbocation, carbocations are stabilized with R-groups.	CH_3>1°>2°	S_N2 rxns are backside attacks, so bulkiness should be minimized.
Nucleophile	no effect	weak nucleophiles are preferred because a carbocation reacts readily. The weak Nuc^- encourages S_N1 rxn by just waiting for carbocation to form.	> for better Nu	Use a good Nuc^- because we need to promote a collision. A properly chosen solvent will help nucleophile strength.
Leaving Group	little effect	To promote formation of carbocation, utilize best LG possible.	no effect	Remember that HO^- and RO^- never leave. Stabilized neg. charges by size or resonance make good LG's.
Solvent	> in polar protic	Again, the H-bonding afforded by polar protic solvents (H_2O, ROH, NH_3) stabilize the carbocation, thus encouraging S_N1.	> in polar aprotic	Protic solvents reduce nucleophile strength because of solvation, so the aprotic solvent makes the Nuc^- less stable and, hence, more reactive. The "polar" aspect is needed for dissolution of Nuc^-.

Lesson 26.1: The E2 Mechanism

E2 stands for "Elimination bimolecular" and, like S_N2, the E2 reaction is a one-step, concerted process. The elimination means that a small molecule is eliminated from the substrate molecule. The general E2 mechanism is shown.

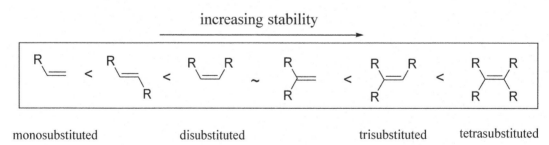

HX has been eliminated from the starting RX

The "2" means the rate-determining step involves the collision of two species, so this requires a bimolecular rate law:

$$\text{Rate of E2} = k[\text{R--LG}][\text{B}^-]$$

Notice in the E2 mechanism that there are three curved arrows: one showing the base taking a proton (H^+), one showing the H-atom's former bonding electrons falling into a double bond, and the third arrow showing the halogen leaving with its bonding electrons. Also note that we use the term "base" when discussing an elimination reaction. A hydroxide ion, HO^-, acts as nucleophile in a substitution reaction (where it donates electrons to an electrophilic C), but as a base in an elimination reaction (where it deprotonates the substrate).

Since the product of an elimination reaction is an alkene we should be aware that alkenes have different relative stabilities based on the groups that are on the C=C carbons. Stability of the alkene affects the outcome of the elimination product, which we'll see soon.

The more substituents on C=C carbons, the more stable.

increasing stability

monosubstituted   disubstituted   trisubstituted   tetrasubstituted

E2 and S_N2 reactions are competing reactions since they occur under the same conditions. Strong bases, like HO^- and RO^-, and high temperatures favor E2 reactions. The E2 is an example of a "β-elimination" because the beta-hydrogen atom is the one removed by the base:

Br
$CH_3-CH-CH_3$
α
β

"alpha" to the Br
"beta" to the Br

$CH_3CH_2-\overset{CH_3}{\underset{CH_3}{C}}-Br$ β

The major products of an E2 reaction can be generally predicted by **Zaitsev's Rule**, which states that the more substituted alkene is formed by removing a proton (H^+) from the β-carbon with the fewest H-atoms. This leads to the most-substituted, most stable alkene as the product.

Predict the major elimination product.

$\xrightarrow[\Delta]{HO^-}$

Br

β

implies

only one H

CH_3
$H_3C-CH-CH_3$
CH
Br

three H-atoms

so these electrons fall in towards the leaving Br⁻

H

Br

→

major product

The rate of the E2 reaction can also be influenced by leaving group identity, substrate substitution degree, base strength, and reaction solvent. The better the leaving group (the more stable the species displaced from the substrate), the faster the reaction will be because there is a lower activation barrier to leaving group departure.

Recall that more non-H substituents on the carbons of a C=C bond makes the alkene more stable. Because the C with the halide leaving group on it ends up in the C=C bond, more non-H substituents at that site leads to more stable products. So, **in terms of E2 rate: methyl halide < 1° alkyl halide < 2° alkyl halide < 3° alkyl halide (fastest)**. This is opposite to that of the S_N2 reaction rates.

Fastest   3°   2°   1°X CH_3X Slowest
               X   X

Increasing E2 reaction rate   no β carbon

A strong base is required for the E2 reaction, and the stronger the base, the faster the E2 reaction. Because the solvent influences base strength, we must choose a solvent that solvates the reagents without diminishing the base strength. As discussed in the substitution reactions section, this means that **the E2 rate is fastest in polar aprotic solvents**.

The major products of an E2 reaction are typically predicted by Zaitsev's Rule. If we don't want the Zaitsev product we can exploit steric hindrance between the base and the substrate, making it more difficult for the base to abstract the H^+ ion. The non-Zaitsev product is called the Hofmann product.

Steric hindrance between a bulky base and substrate will generally be greater for deprotonation of a more substituted site (below left) than the less substituted site on the substrate (below right):

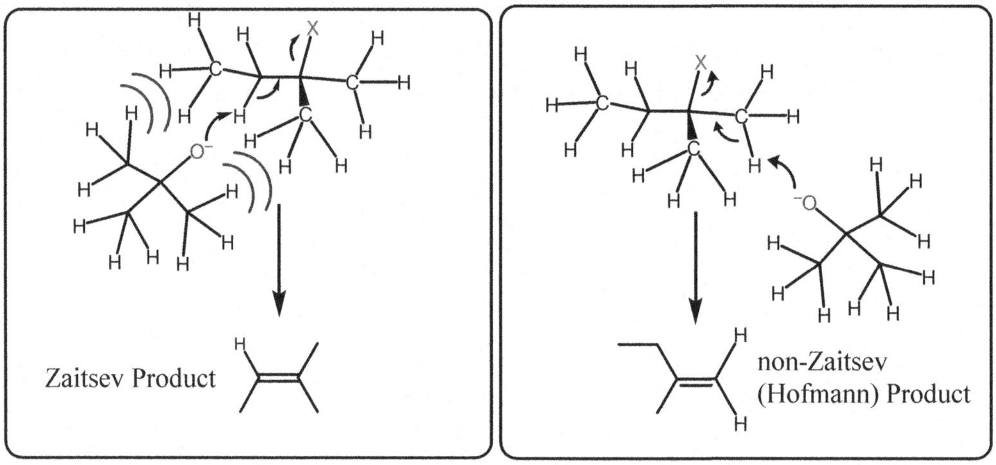

Zaitsev Product

non-Zaitsev (Hofmann) Product

The take home Lesson from this is that **as the steric encumbrance of the base increases, we will get a larger percentage of the Hofmann product** in an E2 reaction.

Example 26.1

Which base will lead to a higher amount of 1-hexene upon reaction with 2-bromohexane: NaOH or $NaOCH(CH_3)_2$ (sodium isopropoxide)?

Solution 26.1

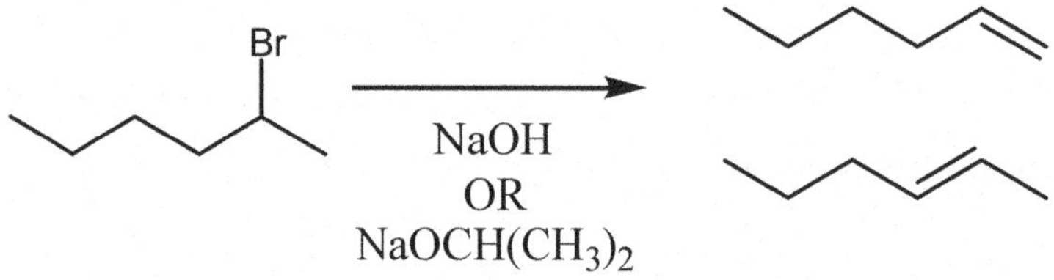

The isopropoxide anion is a much bulkier base than is hydroxide, so it will have a greater steric encumbrance to deprotonate the more substituted site needed to access 2-hexene. A greater yield of the non-Zaitsev product 1-hexene will thus be formed when sodium isopropoxide is added as the base.

As you can probably conclude by now, E1 stands for "elimination unimolecular". As was the case for the S_N1 reaction, formation of a carbocation is the rate-determining step in the E1 mechanism. A general mechanism for an E1 reaction, rate law, and energy diagram are shown here:

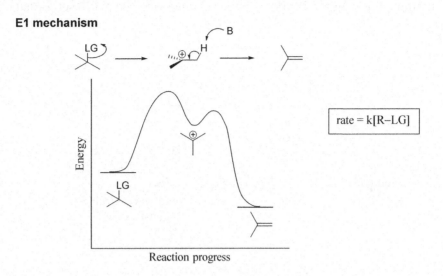

The rate of the E1 reaction can also be influenced by the leaving group identity, substrate substitution degree, and the reaction solvent. The better the leaving group (the more stable the species displaced from the substrate), the faster the reaction will be because there is a lower activation barrier to leaving group departure.

In the E1 mechanism, carbocation formation is rate-limiting. More stable carbocations form faster. So, in terms of rate: methyl halide < 1° alkyl halide < 2° alkyl halide < 3° alkyl halide (fastest). Indeed, **the carbocations that would form from methyl and 1° alkyl halides would be so unstable that methyl and 1° halides will never undergo E1 reactions.**

Because the solvent influences the stability of ions, we must choose a solvent that stabilizes both the carbocation and leaving group formed in the rate-limiting step. As discussed in Lesson 20, polar protic solvents provide the strongest intermolecular forces with ions. For this reason, the **E1 reaction is fastest in polar protic solvents.**

Lesson 26.3. The E1 Mechanism and Alcohols

As a reminder, alcohols have a bad leaving group (an OH) that is readily converted to a good leaving group when it is protonated by an acid. When alcohol undergoes elimination, it is called a dehydration because the small molecule eliminated is a H_2O molecule. Secondary (2°) and tertiary (3°) alcohols undergo E1 elimination and 1° alcohols undergo the E2 reaction. The acid used is usually H_2SO_4, though H_3PO_4 also works.

protonation   LG leaves   proton removal
                                         Zaitsev's rule

Lesson 27. Oxidation and Reduction: Definitions

In organic chemistry, we use a simplified definition of oxidation and reduction, with a focus on the carbon atoms in a structure. In the simplest definition, oxidation is defined as a reaction leading to more C–O bonds and/or fewer C–H bonds, whereas reduction is defined as a reaction leading to fewer C–O bonds and/or more C–H bonds. For a broader definition, we would substitute "C–O bonds" with "C–EN" bonds, where "EN" is any element more electronegative than C.

<u>Example 27.1</u>

Label each of the following reactions as being an oxidation, a reduction or neither:

A) 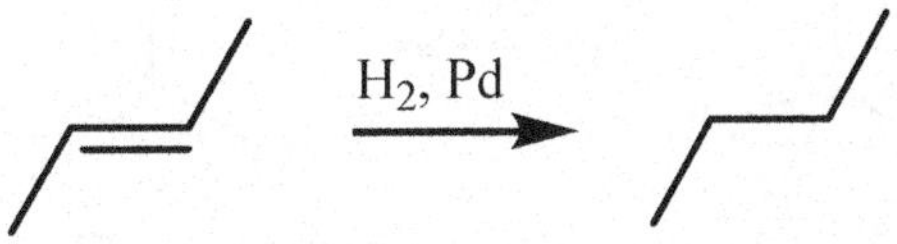
H2, Pd

B) 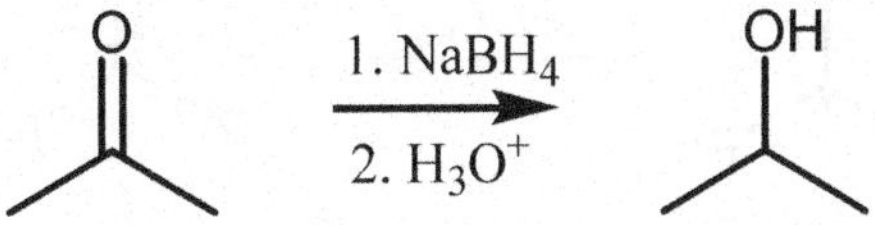
O
1. NaBH4
2. H3O+
OH

C) 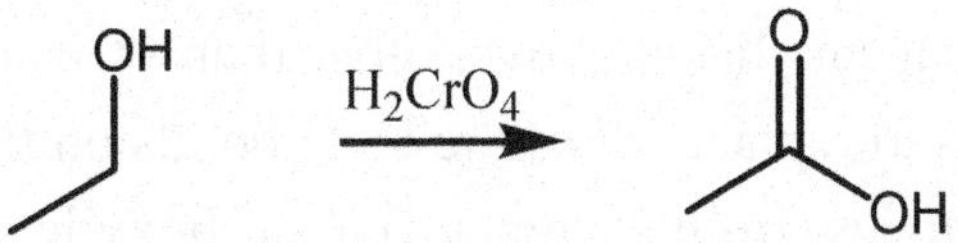
OH
H2CrO4
O
OH

D) 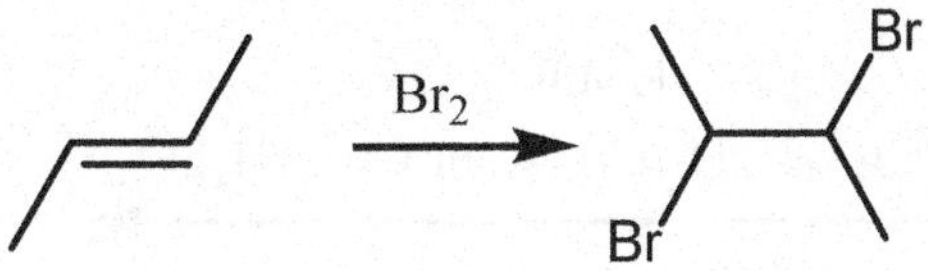
Br2
Br
Br

<u>Solution 27.1</u>

Reaction A: has more C–H bonds in the product than in the reactant, so this is a reduction. Reaction B has fewer C–O bonds in the product than in the reactant, so this is a reduction. Reaction C has more C–O bonds in the product than in the reactant, so this is an oxidation. Reaction D has more C–EN bonds in the product than in the reactant, so this is an oxidation.

Lesson 28. Reactions of Alcohols: Oxidation

The carbon in an alcohol to which the OH group is attached (called the "carbinol" carbon) can be oxidized by several chromium reagents, typically with added acid. Those that we will cover in this book are: H^+/CrO_4^{2-}, $H^+/Cr_2O_7^{2-}$, CrO_3/H_2SO_4 (reagents for what is called the "Jones Oxidation"), $KMnO_4$ and pyridinium chlorochromate (abbreviated PCC).

PCC is a reagent that can replace one C–H bond on the carbinol with another bond to the O of the OH group (which must be accompanied by loss of H from the OH group). This leads to the formation of an aldehyde (from 1° alcohol) or ketone (from 2° alcohol):

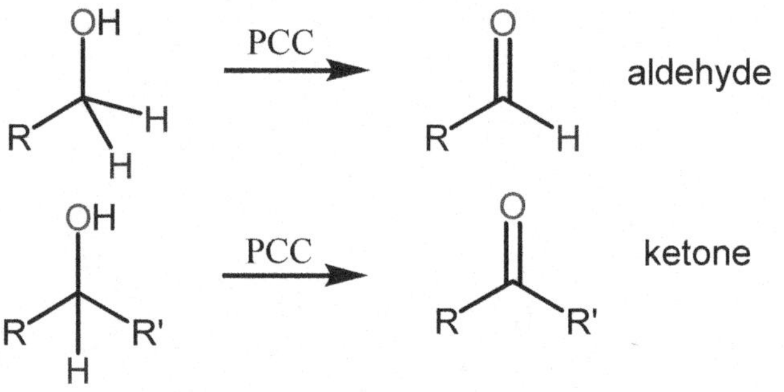

All of the chromium reagents listed above, other than PCC, are more powerful oxidants. They are sufficiently strong oxidants to replace *all* of the C–H bonds on the carbinol with bonds to O. This will lead to the formation of ketones (from 2° alcohol) or carboxylic acids (from 1° alcohol):

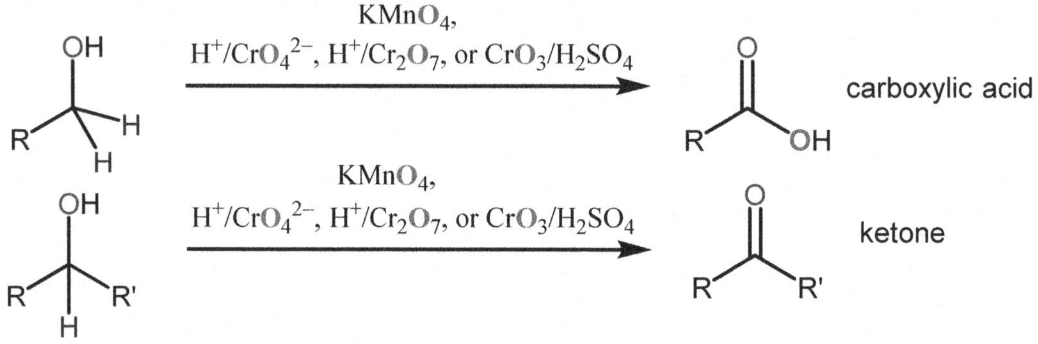

Note that, in the case of 2° alcohols, reaction with PCC or the more powerful oxidizing agents both afford ketones because there is only one H on the carbinol that can be changed to a C–O bond. However, when the reactant is a 1° alcohol, PCC will yield different products than Jones Oxidation.

Example 28.1

Provide the major product for each of the following reactions:

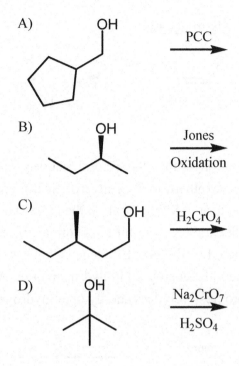

A) [cyclopentylmethanol] OH → PCC →

B) [2-butanol] OH → Jones Oxidation →

C) OH → H₂CrO₄ →

D) OH → Na₂CrO₇ / H₂SO₄ →

Solution 28.1

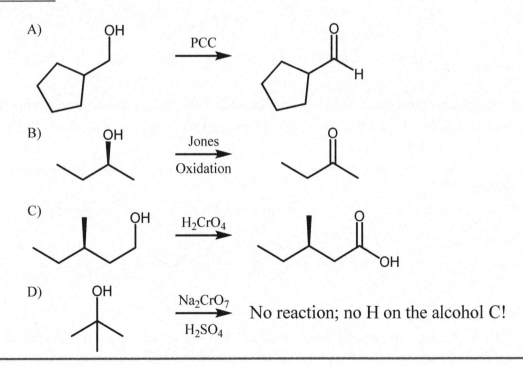

A) OH → PCC → [cyclopentanecarbaldehyde]

B) OH → Jones Oxidation → [butanone]

C) OH → H₂CrO₄ → carboxylic acid OH

D) OH → Na₂CrO₇ / H₂SO₄ → No reaction; no H on the alcohol C!

103

Lesson 29.1. Stability of Alkenes and Heats of Hydrogenation

Hydrogenation of an alkene leads to addition of an H atom to each C atom in the C=C bond:

Hydrogenation

The resulting σ-bonds are stronger than the π-bond that was replaced. As a result, hydrogenation of an alkene to an alkane will be exothermic (i.e., give off heat), and this is referred to as the "heat of hydrogenation". The lower the heat of hydrogenation the more stable the alkene.

As we saw in Lesson 26, the ranking of stability of simple alkenes is tetrasubstituted > trisubstituted > disubstituted > monosubstituted > ethylene. Ethylene is the least stable of the alkenes (i.e., highest in energy), so it will give off the most energy upon hydrogenation. As an alkene becomes progressively more stable, it will give off progressively less energy upon hydrogenation.

Degree of substitution:	none	mono	di			tri	tetra
	$H_2C=CH_2$						
ΔH_h: (kcal/mol)	−32.8	−30.1	−28.7	−28.5	−27.5	−27.0	−26.6

In the series above, the heats of hydrogenation follow the order *cis*-2-butene > 2-methylpropene > *trans*-2-butene. The reason that the *cis*-isomer is less stable than the *trans*-isomer is that the *cis*-isomer has unfavorable steric interactions.

Example 29.1

Which of the following alkenes would have the higher heat of hydrogenation?

Solution 29.1

Both molecules are disubstituted alkenes, but in cyclohexene the double bond has *cis* stereochemistry, whereas the molecule on the right (*trans*-2-butene) has *trans* stereochemistry.

Cyclohexene is less stable (i.e., higher in energy) than *trans*-2-butene due to steric interactions and will thus release more energy upon hydrogenation of the C=C bond. Cyclohexene will have the higher heat of hydrogenation.

Lesson 29.2. Hydrogenation of Alkenes to form Alkanes

The reaction involving adding two hydrogens to an alkene to make an alkane is called "catalytic hydrogentation". The typical conditions used to perform alkene hydrogenation reactions are H_2 (or D_2) in the presence of a transition metal (Ni, Pd, or Pt) as a catalyst. Under these conditions, a molecule of H_2 first reacts with the metal surface, then the alkene reacts via a series of steps shown below to form an alkane:

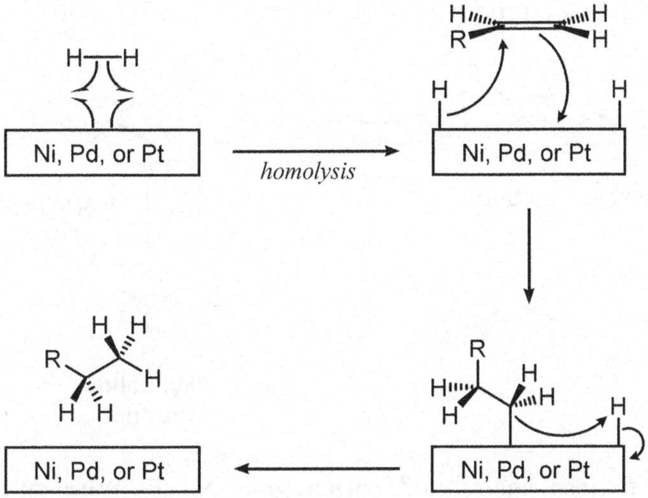

Because both H atoms added to the alkene come from the metal surface, both H atoms will add to the same face of the C=C bond (i.e., they will always add *syn*).

Example 29.2

Draw the major product of the reaction shown below:

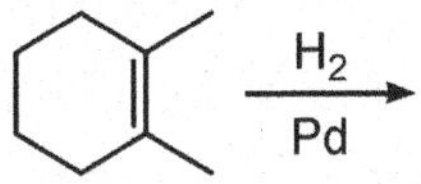

Solution 29.2

The hydrogens add *syn* with respect to each other, which restricts the 2 methyl groups to be *syn* with respect to each other and the product of this reaction, is a meso compound, so only one stereoisomer is possible.

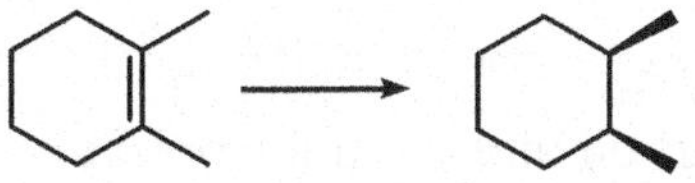

105

Lesson 30. Reactions of Alkenes I: Hydrohalogenation and Hydration

Lesson 30.1. Hydrohalogenation – Electrophilic Addition by H^+, then nucleophilic Attack by X^-

Hydrohalogenation refers to the addition of H–X (X = Cl, Br, I) across a C=C bond. H–X is a strong acid, so it will dissociate to H^+ and X^- in solution. Thus, H^+ is the electrophile that reacts with the C=C bond at the less substituted carbon to afford a carbocation at the more substituted carbon. Once the most stable carbocation has been formed, X^- will attack the empty *p*-orbital on the sp^2-hybridized carbon. When there is no difference in energy between X^- approaching the carbocation from above vs. below the plane, which is generally the case, then the reaction products will be derived equally from both approaches. If the product is a chiral molecule where the carbon center bonding to the halide is a chiral center, then two stereoisomers are produced (one wedge and one dash on C—X).

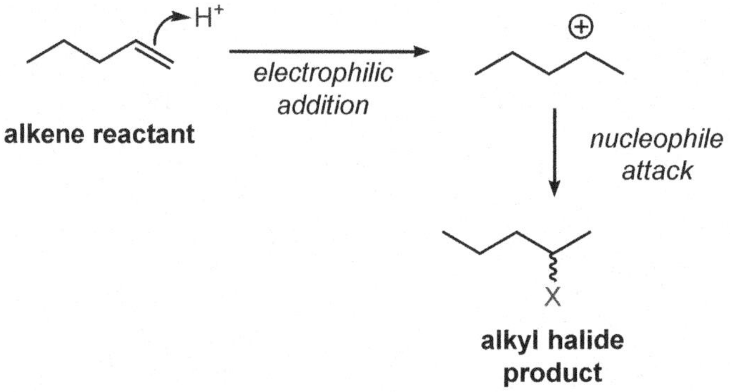

The addition of H^+ to less substituted carbon and X^- to more substituted carbon is called "Markovnikov's Rule"

Markovnikov's rule states that, for the addition of a given molecule X–Y to an alkene, the more electronegative element ends up on the more substituted carbon.

Because there are two carbons in a C=C bond, electrophilic addition to an alkene followed by nucleophile attack could theoretically afford two products that differ in relative arrangement of electrophile and nucleophile (i.e., E on carbon 1 and Nu on carbon 2, or vice versa), and these are termed "regioisomers". However, the addition of E^+ to an alkene affords a carbocation intermediate, so E^+ will add to the less substituted carbon, leaving the more stable cation on the more substituted carbon, as illustrated in the electrophilic addition of E^+ to 1-butene:

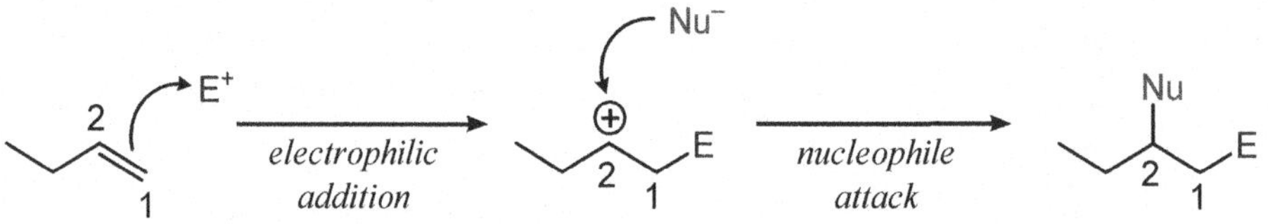

Subsequent attack of Nu^- to this carbocation affords the major product, wherein the electrophile has attached to the less substituted carbon and the nucleophile to the more substituted carbon. Because the reaction inherently favors one regioisomer, the reaction is said to be "regioselective".

Example 30.1

Draw the major product of the reaction shown in the box:

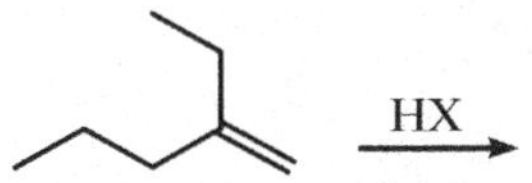

Solution 30.1

In the first step of the reaction, the electrophile H^+ adds to the π-bond and forms a C–H bond at the less substituted alkene carbon, which allows the carbocation to form at the more substituted carbon. After the carbocation has formed, X^- will then attack the empty p-orbital from either above or below the sp^2-hybridized carbon's plane. Because there is no difference in energy between these two approaches, a 1:1 mixture of the two enantiomers will be formed.

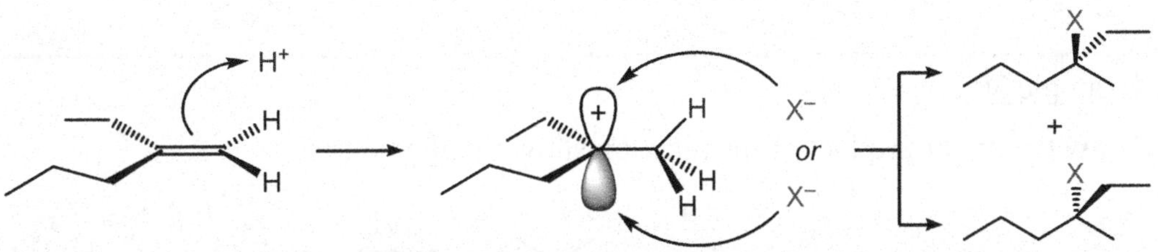

Lesson 30.2. Hydration – Electrophilic Addition by H^+, then Nucleophilic Attack by H_2O

Hydration refers to the addition of H–OH (i.e., water) across a C=C bond. However, H_2O is a weak acid and is not electrophilic enough to react with the C=C bond. Thus, hydration reactions require the addition of a catalytic amount of a strong acid. HX (X = Cl, Br, I) is not used because then the highly nucleophilic X^- would be the nucleophile and the result would be hydrohalogenation, not hydration. The most commonly-used acid for alkene hydration reactions is H_2SO_4. As before, H^+ is the electrophile that reacts with the C=C bond to form a carbocation. Then the water molecule attacks the carbocations and forms a new σ bond. The product is a cationic molecule containing a protonated OH group, which is then deprotonated by HSO_4^- to afford an alcohol and regenerate the initial H_2SO_4 (which is why a catalytic amount of acid is sufficient and a stoichiometric amount is unnecessary). Hydration reaction also follows "Markovnikov rule", where the H^+ adds to the less substituted carbon and the alcohol adds to the more substituted carbon. The mechanism for the hydration reaction is given in the figure below.

The reaction mechanism shows: alkene + H⁺ → *electrophilic addition* → carbocation intermediate; then *Nucleophilic attack* with water (X⁻ shown) → *oxonium intermediate* → *−H⁺ deprotonation* → alcohol product.

X⁻ = (hydrogen sulfate ion, HSO_4^-):

$$O=\overset{\overset{\displaystyle O^-}{|}}{\underset{\underset{\displaystyle OH}{|}}{S}}=O$$

Example 30.2

Draw the major product of the reaction shown in the box:

$$\text{(but-1-ene)} \xrightarrow[\text{H}^+]{\text{H}_2\text{O}}$$

Solution 30.2

In the first step of the reaction, the electrophile H^+ adds to the π-bond and forms a C–H bond at the less substituted alkene carbon, which allows the carbocation to form at the more substituted carbon. After the carbocation has formed, water will then attack the carbocation and the product is a 1:1 mixture of the two enantiomers.

$$\xrightarrow[\text{H}^+]{\text{H}_2\text{O}}$$

(butan-2-ol, as a pair of enantiomers)

Lesson 31. Reactions of Alkenes II: Halogenation

Lesson 31.1. The Halonium Intermediate

The proper mechanism for halonium formation does not involve the reaction of a C=C bond with free X^+, but rather with polar or polarizable X_2. Upon encountering the halogen, π-electrons from the C=C bond attack $X^{\delta+}$ while *simultaneously*, a lone pair on X coordinates to the more substituted carbon in the C=C bond, with release of a halide anion. Note that the halonium can form on either face of the C=C unit and, if there is no energy difference between X–X approaching the C=C bond from above or below, then a 50:50 mixture of both stereoisomers will be formed:

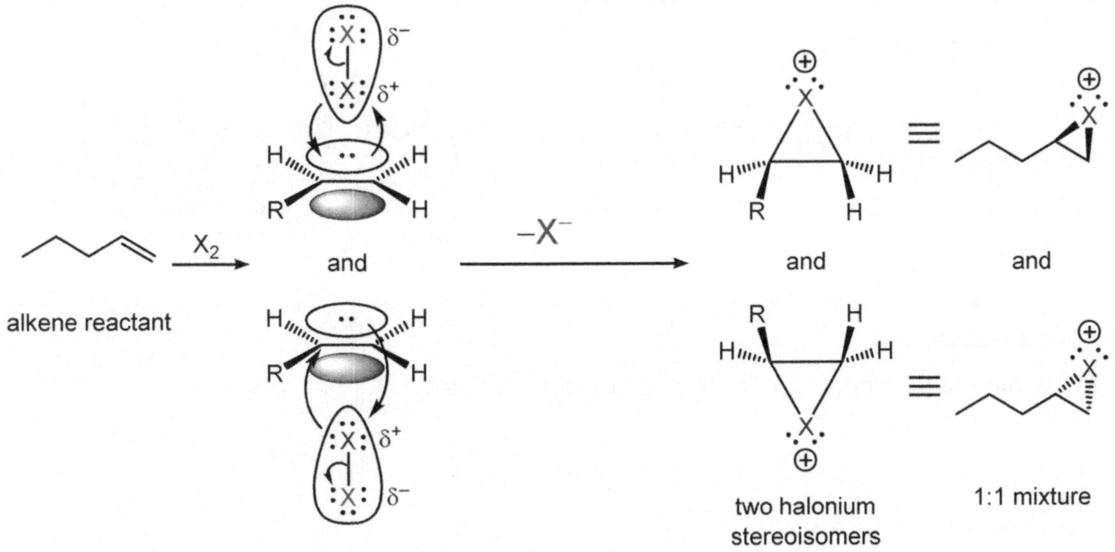

Example 31.1

How many stereochemically unique halonium isomers will form in the reaction shown below?

Solution 31.1

The bromonium intermediate is formed when one of the Br atoms in Br_2 adds to one face of the alkene π-bond. As always, this addition can occur either above or below the alkene plane and a mixture of stereoisomers will be produced. With *cis*-2-butene as the reactant, however, the bromonium intermediate is a meso compound: although there are two chiral carbons, there is a mirror plane bisecting the C–Br–C angle, so the molecule is a meso compound and is achiral. Thus, there is only one possible stereochemically unique isomer that will form in this reaction.

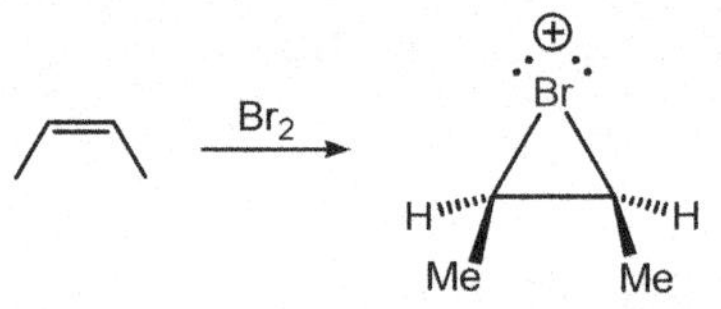

Lesson 31.2. Halogenation – Nucleophilic Addition of X⁻ to the Halonium Intermediate

The halonium intermediate reacts with the halide nucleophile in S_N2-like mechanism with inversion of configuration. When this transformation is complete, the two X atoms end up on opposite sides of what was the C=C bond with respect to each other (they are *anti*).

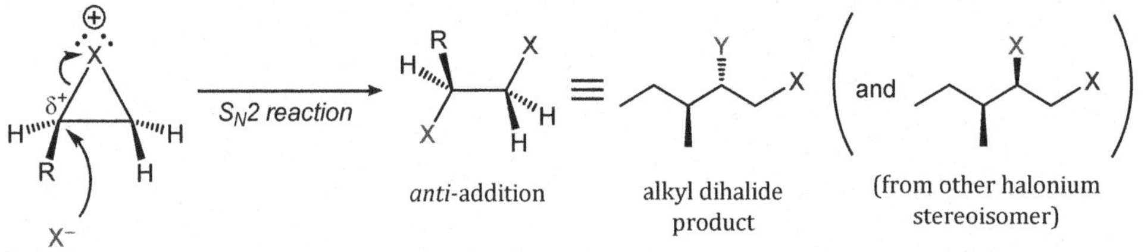

anti-addition   alkyl dihalide product   (from other halonium stereoisomer)

Example 31.2

Draw the major product(s) of the reaction shown below:

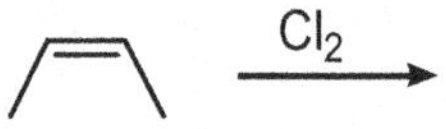

Solution 31.2

The chloronium intermediate is formed when one of the Cl atoms in Cl_2 adds to one face of the alkene π-bond. As always, this addition can occur either above or below the alkene plane and a mixture of stereoisomers will be produced. With *cis*-2-butene as the reactant, however, the chloronium intermediate is a meso compound. Then the chloride ion will add in an *anti* fashion to either carbons of the chloronium intermediate since they are both of same degree of substitution, giving two products in total:

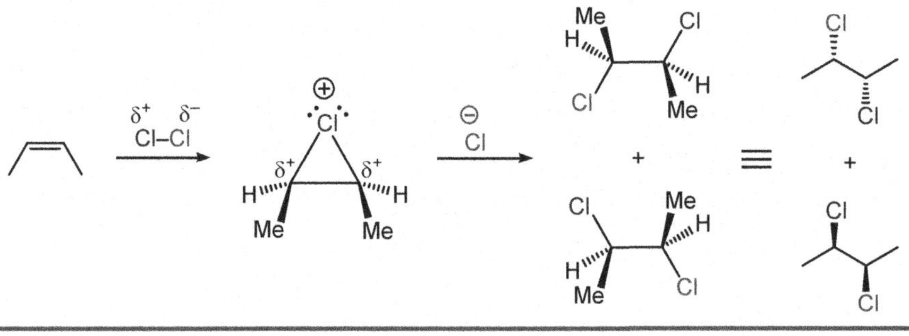

Lesson 32. Reactions of Alkenes IV: Hydroboration/Oxidation

Lesson 32.1. Hydroboration – Concerted Addition of $B^{\delta+}-H^{\delta-}$ to a C=C Bond

The alkene reactions we have studied so far have given us Markovnikov products, but there are instances in which it might be synthetically useful to be able to access an anti-Markovnikov product. Say, for example, we wish to make an anti-Markovnikov alcohol. The reagent to use in such a scenario is BH_3 (which is occasionally written as B_2H_6) because its bonds are polarized as $B^{\delta+}-H^{\delta-}$ (H is more electronegative than B). As a result, the B–H bond will align over the C=C bond such that the $H^{\delta-}$ is above the $\delta+$ of the more substituted carbon and the $B^{\delta+}$ is above the less substituted carbon. The π-electrons of the C=C bond then attack the $B^{\delta+}$ (which has a strong pull for electrons because it only has 6 valence electrons) to make a C–B bond, while simultaneously, the $H^{\delta-}$ attacks the more substituted carbon with the electrons from the B–H bond to make a C–H bond. This **hydroboration** step is concerted – there are no intermediates of any kind, so no rearrangement can occur, which makes this reaction regioselective. In addition, the H and BH_2 groups end up on the same side of what used to be the C=C bond (the product is *syn*), so this reaction also exhibits stereoselectivity. Keep in mind, though, that the B–H bond can add either above or below the C=C bond, so unless there is a difference in energy between the two approaches, a mixture of stereoisomers will be formed.

Lesson 32.2. Oxidation – Replacing BH$_2$ with OH

If our goal is to make an anti-Markovnikov alcohol, we need to replace the BH_2 group with an OH group, which can be achieved under oxidizing conditions: H_2O_2, NaOH, H_2O. This oxidation occurs via a series of steps that retain stereochemistry at the C:

This series of steps results in the net conversion of BH$_2$ to OH without rearrangement and with retention of stereochemistry (i.e., regioselective and stereoselective). The combination of these two reactions, written as **"hydroboration/oxidation", results in anti-Markovnikov addition of H and OH across a C=C bond with *syn*- stereochemistry.**

Example 32.1

What is the major product of the reaction shown below?

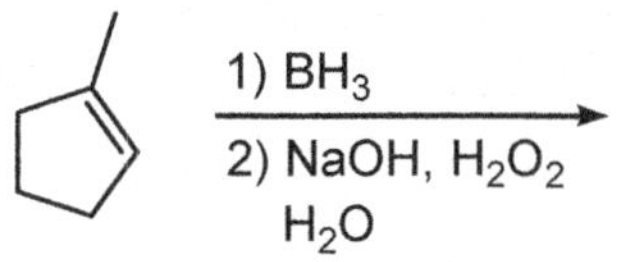

1) BH$_3$
2) NaOH, H$_2$O$_2$
   H$_2$O

Solution 32.1

The reagent BH$_3$ will react with an alkene π-bond to add H to the more substituted alkene carbon and BH$_2$ to the less substituted carbon with *syn* stereochemistry. With 1-methyl-1-cyclopentene as the reactant, BH$_3$ can add from either above the C=C plane, pushing the Me substituent down, or below the C=C plane, pushing the Me substituent up (step *i*). As a result, the first reaction, hydroboration, forms a 1:1 mixture of enantiomers, in which the BH$_2$ and Me groups end up *anti* with respect to each other (because the H, which is usually not drawn, is *syn* with respect to the BH$_2$). The second reaction, oxidation with H$_2$O$_2$ under basic conditions, replaces the BH$_2$ group with an OH group with retention of stereochemistry at the carbon, so each hydroboration product stereoisomer will produce only one alcohol product stereoisomer. Thus, the net hydroboration/oxidation of 1-methyl-1-cyclohexene will afford a 1:1 mixture of an enantiomeric pair of alcohols, in which the OH has been added to the less substituted alkene carbon.

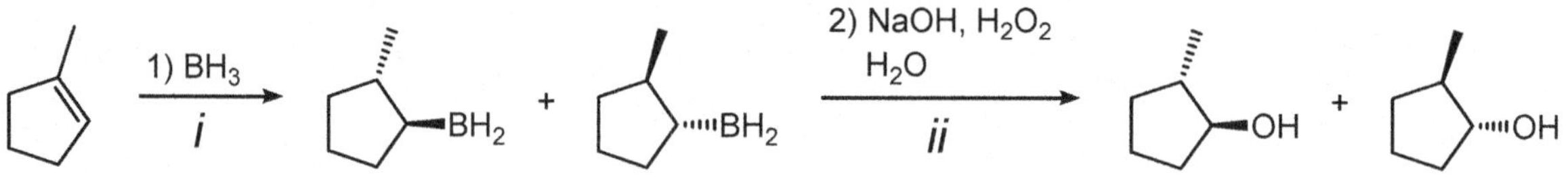

Lesson 33. Reactions of Alkenes III: Epoxidation

Treating an alkene with a peroxyacid, R–C(=O)–O–O–H, will give an epoxide as product and a carboxylic acid as byproduct. The prefix "peroxy-" indicates that there is an O–O single bond present, which is very weak and can be easily broken. A widely used peroxyacid is *m*-chloroperoxybenzoic acid (mCPBA), which contains a peroxycarboxylic acid group.

The mechanism by which mCPBA (or any peroxyacid) mediates the epoxidation is shown here:

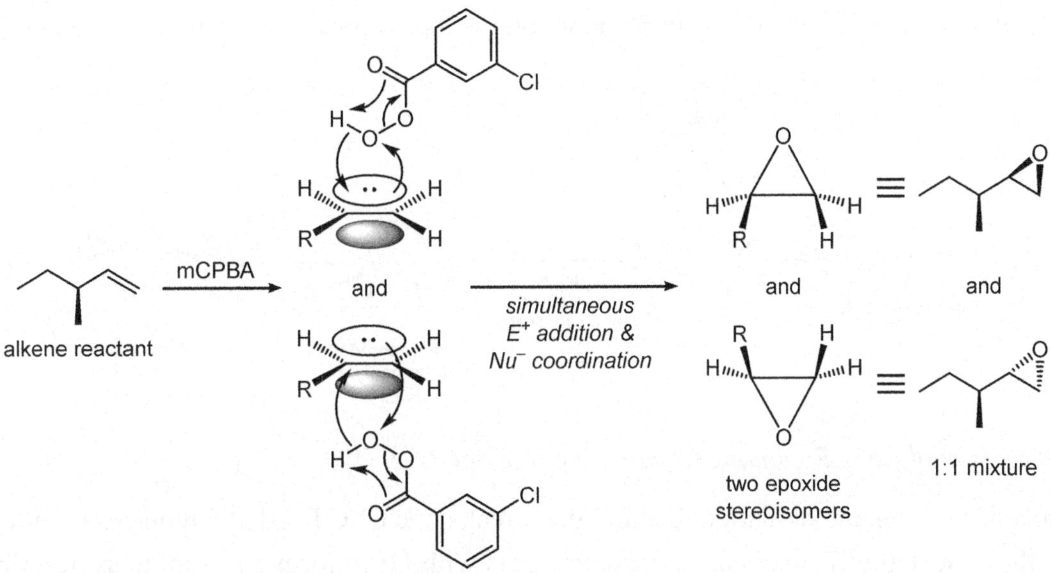

Because all of these transformations occur in a single, concerted step, if we begin the reaction with a *trans*-alkene, those substituents in the epoxide ring will also be *trans* to each other. The reactant mCPBA can react either above or below the plane of the alkene, so a 1:1 mixture of stereoisomers will be produced if there is no difference in energy between these two approaches.

Example 33.1

What is the major product of the reaction shown below?

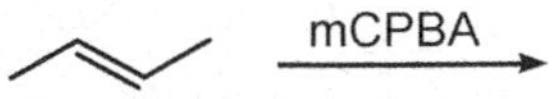

Solution 33.1

The reactant mCPBA functions as a source of "O" atom, which will react with the π-bond in an alkene to make an epoxide. This addition can occur from either above or below the C=C plane, which in the case of *trans*-2-butene will afford a 1:1 mixture of enantiomers.

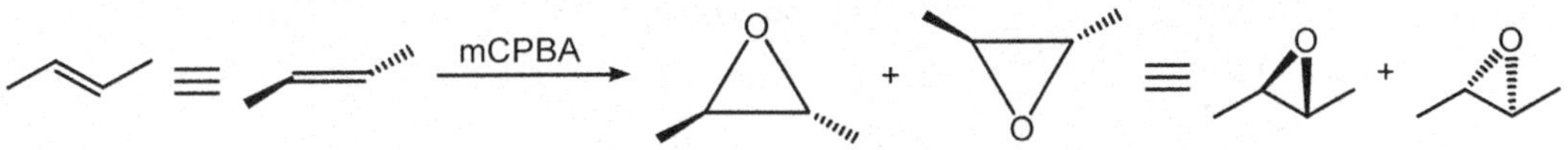

Lesson 34. Reactions of Alkenes V: Ozonolysis

Lesson 34.1. The Ozonide Intermediate

In the Lewis dot structure of ozone (O_3), the very polar O–O bonds (there is a formal +1 charge on the central O) allows the flow of electrons between ozone and an alkene to occur as follows to form an unusual 5-membered ring termed a "molozonide". A molozonide has two O–O single bonds which are very weak, so this molecule will rearrange via a nucleophilic elimination/nucleophilic addition to yield an ozonide, which has only one O–O single bond, and is thus more stable than the molozonide.

Lesson 34.2. Ozonolysis – Replacing C=C with C=O Bonds

Ozonolysis is an alkene oxidation reaction that breaks one C=C bond and replaces it with two C=O bonds. In the first step of ozonolysis, an alkene reacts with O_3 to form an ozonide as described in the previous section. In the second step, the ozonide is subjected to either reducing conditions (common reductants are R_2S or Zn/H_2O) or oxidizing conditions (a common oxidant is H_2O_2). Under reducing conditions, any H atoms attached to the alkene carbons will be retained in the C=O bond-containing products (i.e., as aldehydes). Under oxidizing conditions, any H atoms attached to the alkene carbons will be replaced with OH groups in the C=O bond-containing products (i.e., as carboxylic acids). With a tetrasubstituted starting alkene, both reducing and oxidizing conditions afford the same products (i.e., only ketones will be formed).

Example 34.1

What is the major product of the reaction shown below?

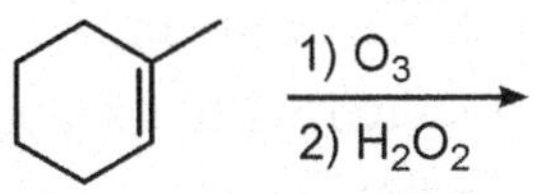

Solution 34.1

The two reactions in this Example have the net effect of cleaving a C=C bond and replacing it with two C=O bonds. Any alkene carbon with two alkyl substituents will be converted to a ketone, regardless of whether oxidizing or reducing conditions are used. Hydrogen substituents, however, will only be retained under reducing conditions, and will be replaced with OH groups under oxidizing conditions. Thus, the oxidation in this example will afford a ketone and a carboxylic acid functional group. If we had instead used reducing conditions, the product would have had a ketone and an aldehyde functional group.

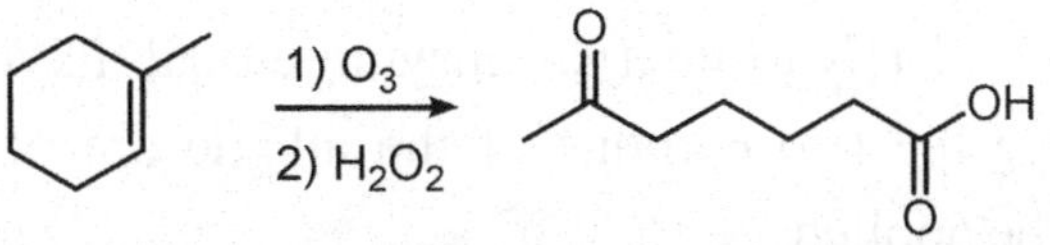

115

Lesson 35. Alkyne Reactions I: Hydrohalogenation

Lesson 35.1. Hydrohalogenation – Electrophilic Addition by H⁺, then Attack by X⁻

When an alkyne is reacted with H–X, H–X undergoes *anti* addition to afford a trans haloalkene. If this trans halo alkene product then is reacted with a second equivalent of H–X, then carbocation will form at the carbon with the X substituent. As a result, the reaction of an alkyne with two equivalents of H–X will result in both X substituents being attached to the same carbon.

Lesson 35.2. Addition of One Equivalent of HX

Adding one equivalence of HX to internal unsymmetrical alkyne will give two products that are *constitutional isomers* since the two carbons of the alkyne are of same degree of substitution and "Markovnikov rule" can't be applied.

Adding one equivalence of HX to internal symmetrical alkyne will give one product since both carbons of the alkyne are symmetrical and the two possible products are identical

As we discussed in Lesson 13, alkynes can be internal or terminal:

116

Whether an alkyne starting material is internal or terminal can have an effect on a reation's outcome. For example, adding one equivalence of HX to terminal alkyne will give one product that follows Markovnikov rule, where the halide X is bonded to the more substituted carbon, the product is called a vinyl halide product.

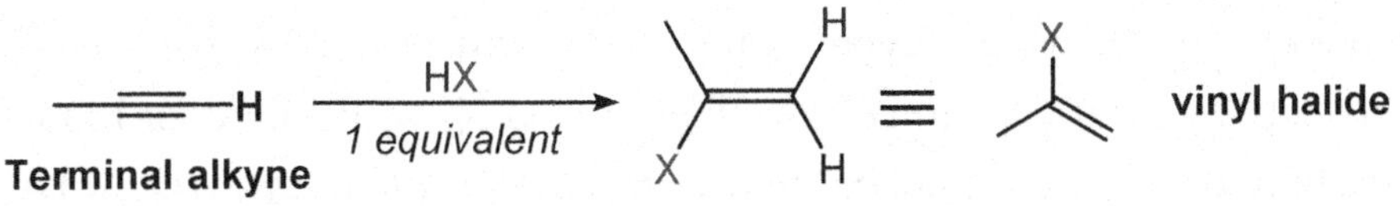

Example 35.1

Draw the major product of the two-step reaction sequence shown below:

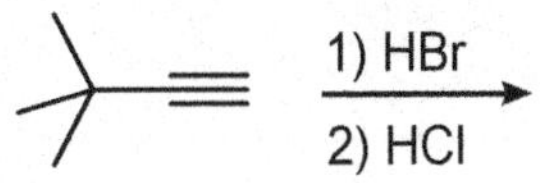

Solution 35.1

The net result of the first reaction is the Markovnikov addition of H and Br to one of the alkyne π-bonds, with *anti* stereochemistry with respect to each other. The second reaction is alkene hydrohalogenation. Briefly, the addition of HCl to the alkene results in the H^+ adding to the less substituted alkene carbon, because the carbocation that forms on the carbon bearing the Br can be stabilized by the lone pairs on Br. Both reactions exhibit Markovnikov regiochemistry, and although the first reaction is stereoselective (*anti* addition), the fact that the second reaction proceeds through an achiral carbocation intermediate means that the net result of the two sequential reactions is not stereoselective.

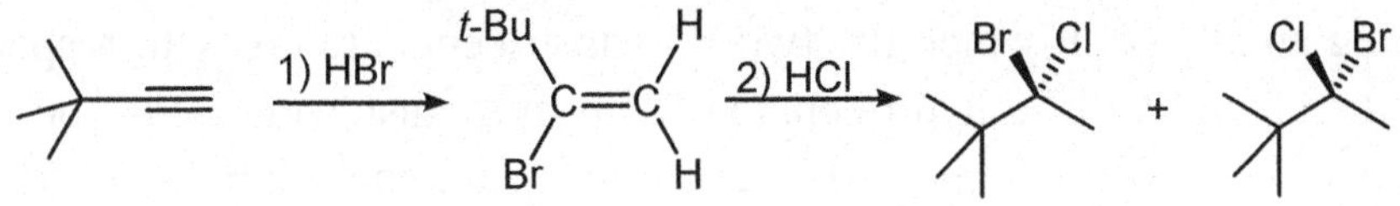

Lesson 36.1. Alkynes to Alkanes

In Lesson 29, we discussed hydrogenation of an alkene to an alkane. If we perform hydrogenation on an alkyne, the first equivalent of H_2 would convert the C≡C bond to a C=C bond. Once formed, the alkene is even more reactive than the alkyne, so the alkene will be rapidly hydrogenated to the alkane. It is not feasible to stop the hydrogenation of the alkyne using H_2 with Ni, Pd, or Pt metal after only one hydrogenation reaction to isolate the alkene. This is termed *exhaustive catalytic hydrogenation*. Alkyne hydrogenation operates via the same mechanism as alkene hydrogenation, so each equivalent of H_2 is added syn across the C≡C bond.

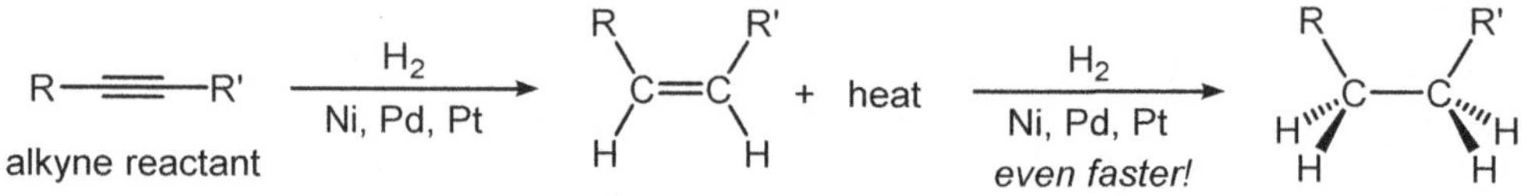

Example 36.1

Draw the major product of the reaction shown below:

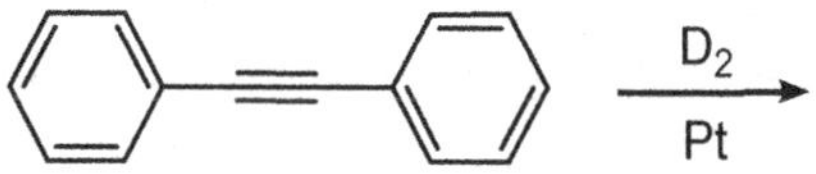

Solution 36.1

Remember that D_2 will exhibit the same chemical reactivity as H_2 because D is just a heavier isotope of H. In the presence of Pt metal, D_2 reacts with the metal surface to break the D–D bond and to make two M–D bonds. When the reactant, diphenylacetylene, binds to the Pt surface, two D atoms transfer sequentially to the same face (i.e., *syn* addition) of one of the π-bonds (step *i*), producing an alkene in which the two Ph substituents are *cis* with respect to each other. As a result, the alkene will be hydrogenated at an even faster rate than the alkyne. Thus, another equivalent of D_2 will undergo *syn* addition to the alkene π-bond (step *ii*), affording an alkane. There is no regioisomerism to consider, because a deuterium is always added to each of the carbons. If the product of the hydrogenation reaction contains chiral centers, then a mixture of stereoisomers will form, because π-bonds can bind to the metal surface from either face. For this Example, however, there is no stereoisomerism that needs to be considered.

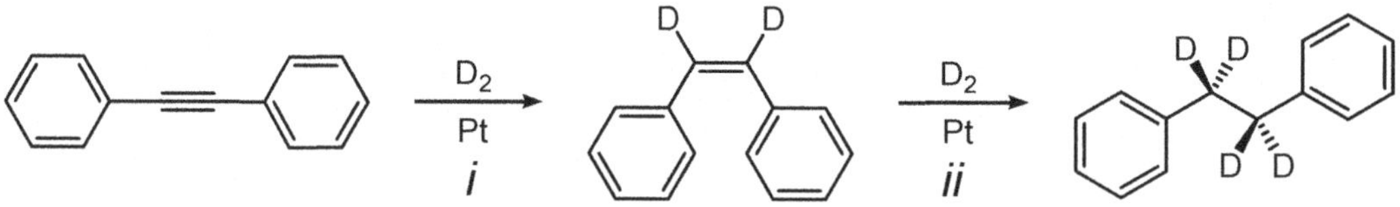

Lesson 36.2. Alkynes to cis-Alkenes

In some instances, it may be useful to have a means to convert an alkyne to an alkene without the hydrogenation proceeding all the way to the alkane. If we can add a compound that reduces the effectiveness of the catalyst (such a compound is called a "poison"), the hydrogenation reaction will proceed more slowly, which will make it possible to stop the reaction from hydrogenating the alkene to alkane. Herbert Lindlar discovered that the addition of lead or sulfur compounds poisoned the Pd catalyst used for hydrogenation, hence the system named "Lindlar's catalyst" will only hydrogenate an alkyne to an alkene. The H_2 is always added *syn* across the C≡C bond.

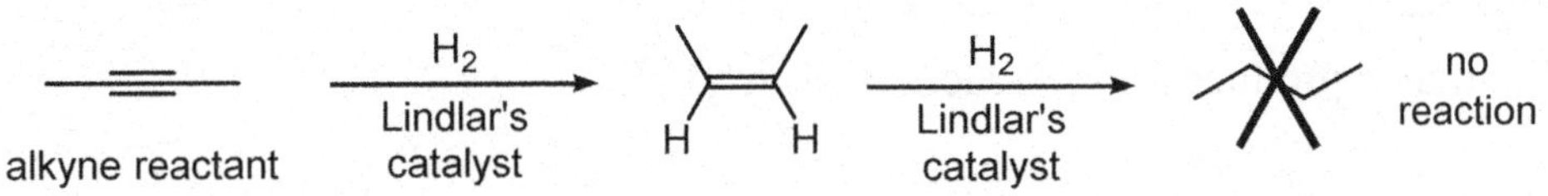

Example 36.2

Draw the major product of the reaction shown below:

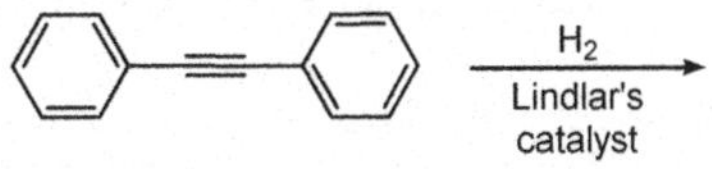

Solution 36.2

H_2 in the presence of Lindlar's catalyst will result in *syn* addition of H atoms to each carbon in the π-bond of diphenyl acetylene to afford a *cis*-alkene.

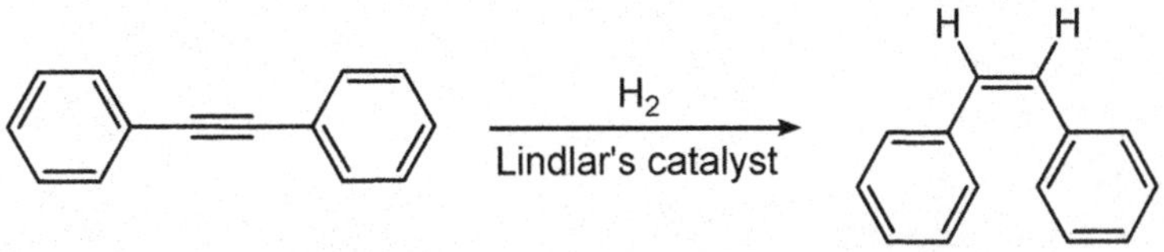

Lesson 36.3. Alkynes to trans-Alkenes

Exposing an alkyne R–C≡C–R' to the conditions of metal sodium or lithium, in liquid ammonia (Na/NH$_3$(l)) will result in stereoselective reaction where the trans alkene will be produced. The mechanism is complex but leads to the anti addition of the two hydrogen atoms across the π bond of the alkyne.

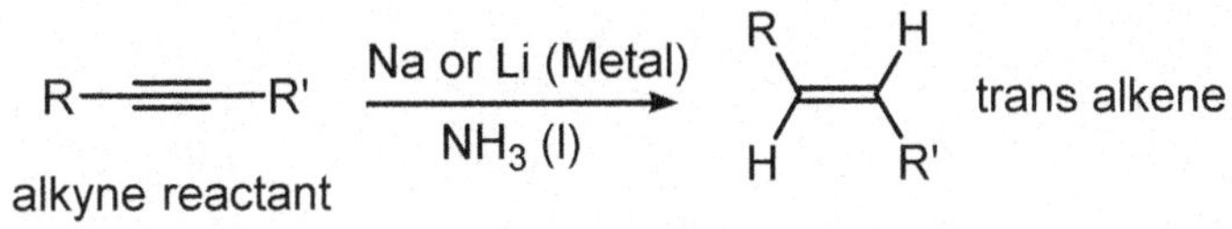

119

If the reacting alkyne is a terminal alkyne, then both H_2/Lindlar's catalyst and Na/NH_3 (l) will result in a terminal alkene where there is no cis or trans present.

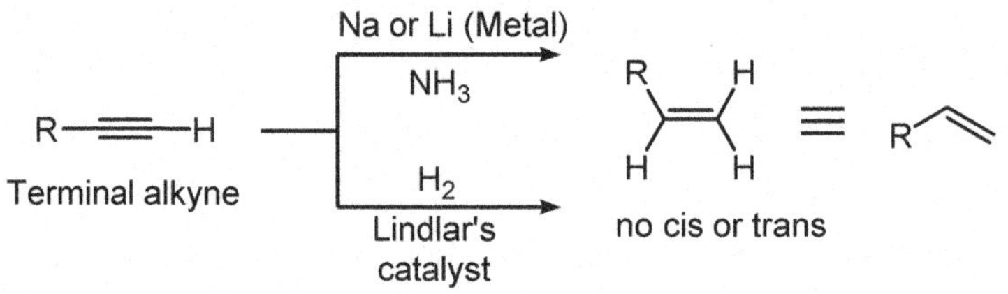

Example 36.3

Draw the major product for the reaction shown below:

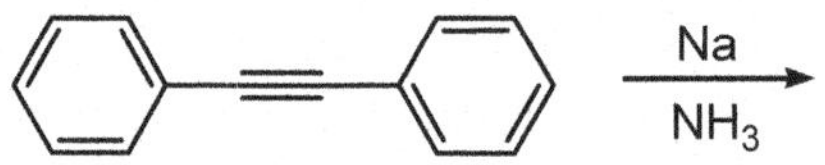

Solution 36.3

Since the reagent is Na/NH_3 and the alkyne is internal, then the trans-stilbene will be the product

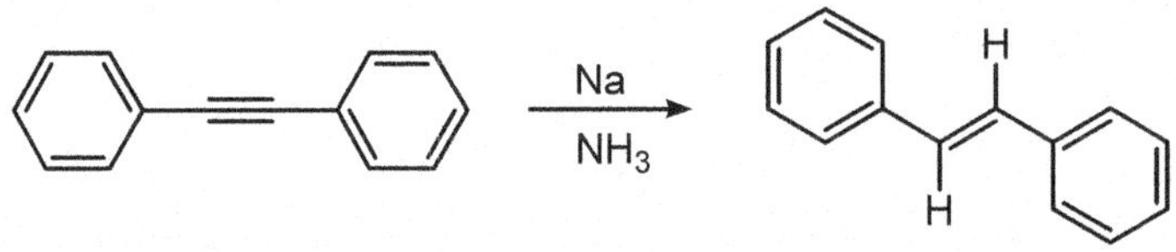

Lesson 37. Alkyne Reactions III: Preparation of Carbonyls

Lesson 37.1. Enol and Keto Tautomers – A Special Class of Constitutional Isomers

The concept of constitutional isomers was introduced in Lesson 9. Constitutional isomers are non-identical molecules with identical molecular formulae, but different bond connectivity, which cannot be interconverted without breaking one or more σ-bonds. In general, σ-bond breakage does not occur readily and requires a significant amount of energy to overcome the activation barrier. However, there are special classes of constitutional isomers that can readily interconvert because the activation barrier to this interconversion is low. Tautomers are one such class, and their constitutional isomerism derives from a C_2H_2O subunit. In the "enol" form, the C_2H_2O subunit comprises an OH group attached to a C=C bond. In the "keto" form, the C_2H_2O subunit comprises a CH_2 group adjacent to a C=O. These two species exist in chemical equilibrium with each other.

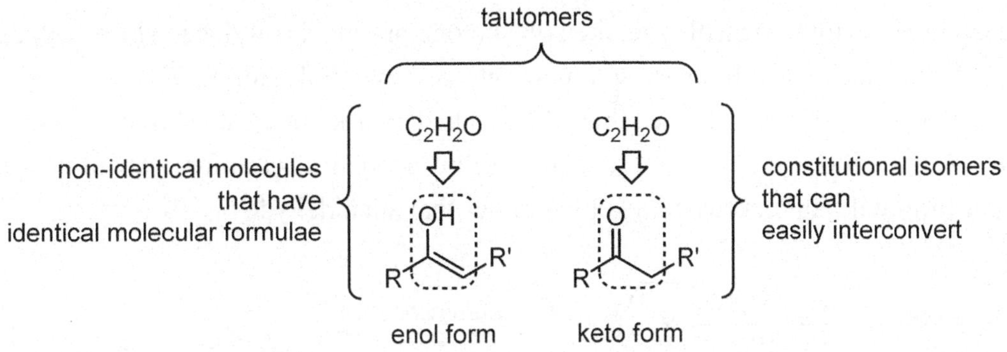

Lesson 37.2. Hydration of Alkynes: Markovnikov Formation of Enols

One difference in typical alkyne hydration reactions with H_2O and H_2SO_4 is that $HgSO_4$ is generally used since alkynes are less reactive than alkenes, this reaction is termed as "oxymercuration". Furthermore, the reactivity of an alkyne proceeds via a different mechanism than an alkene, thus the reaction does not go through a carbocation intermediate.

The product of alkyne hydration is an enol (Enol stands for a vinylic alcohol group), enol product is unstable and it will undergo a change called "tautomerization" to give carbonyl as the constitutional isomer (keto form). The general reaction is depicted below:

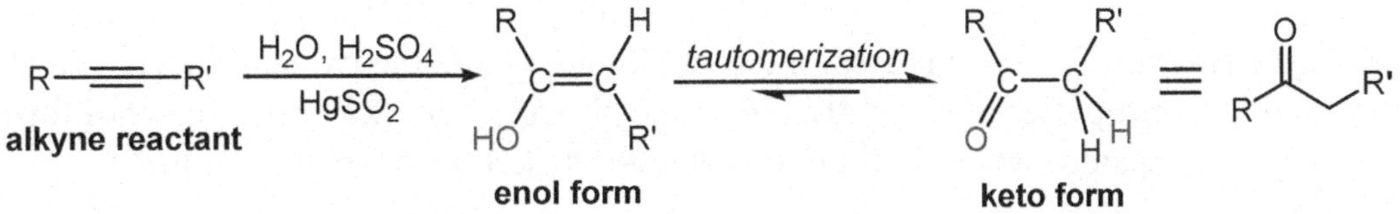

The oxymercuration of an alkyne, formation of an enol, is both regioselective (Markovnikov) and stereoselective (*anti*). The net result of hydration is the conversion of an alkyne to a ketone with Markovnikov regioselectivity. So, if the reactant is an internal unsymmetrical alkyne, the reaction will produce two ketones as constitutional isomers, since both carbons of the alkyne are of same degree of substitution and Markovnikov rule doesn't apply, as shown in this reaction:

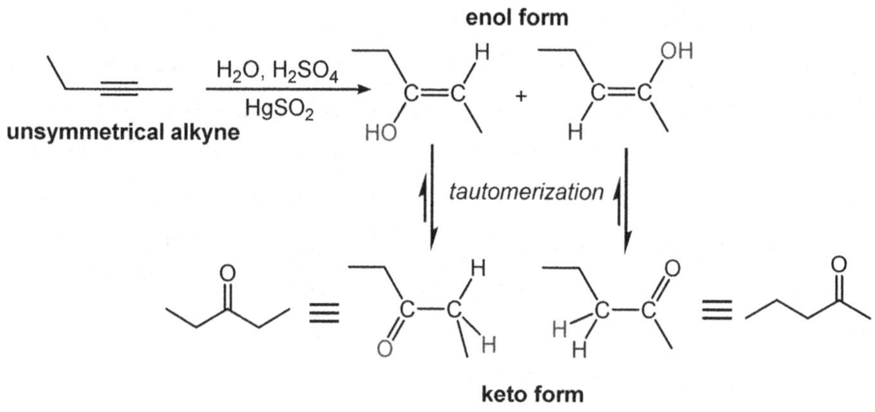

If the reactant is a symmetrical alkyne, then one ketone product is produced since both carbons have same degree of substitution and both are symmetrical, as shown in Example 37.1

If the reactant is a terminal alkyne, then the product is a ketone rather than an aldehyde since, according to Markovnikov rule, and the alcohol will be bonded to the more substituted carbon, as a result the enol form will tautomerize to a ketone rather than an aldehyde.

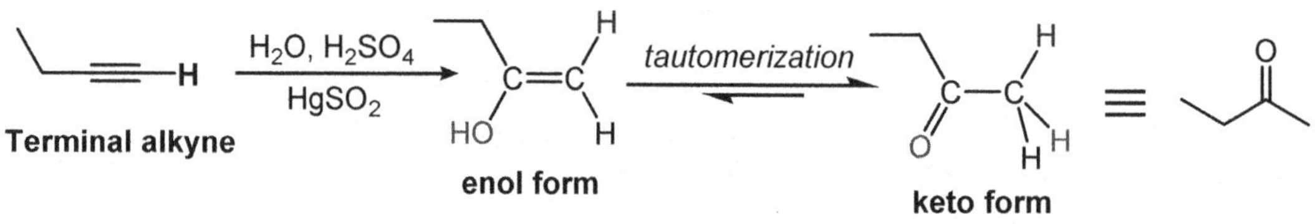

Example 37.1

Draw the major product of the reaction shown below:

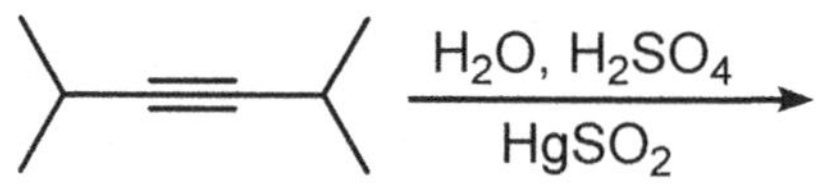

Solution 37.1

The alkyne reactant is a symmetrical internal alkyne, where the two carbons of the alkyne are connected to isopropyl groups. Hydration of the above alkyne will give one enol form (since both carbons are same) which will tautomerize to the ketone as major product

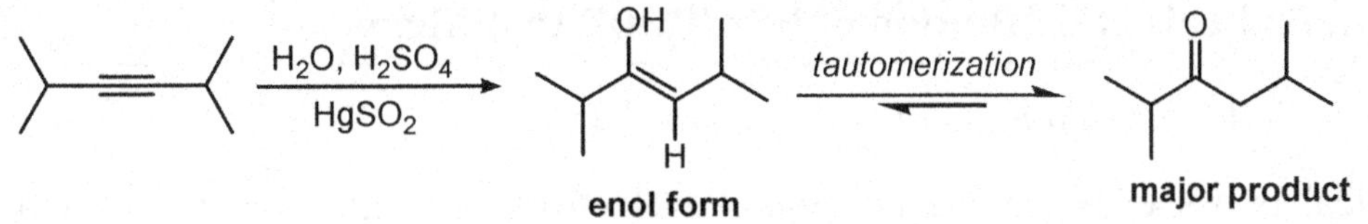

enol form   major product

Lesson 37.3. Hydroboration/Oxidation: Anti-Markovnikov Enol Formation

Hydroboration/oxidation of alkynes proceeds via a mechanism very similar to that of alkenes. In the first reaction (hydroboration), the $B^{\delta+}–H^{\delta-}$ bond adds across the C≡C bond such that the H ends up on the more substituted alkyne carbon and the BH_2 ends up on the less substituted carbon. Because this all happens in a single concerted step, the H and BH_2 end up *syn* with respect to each other. In the second reaction (oxidation), the BH_2 is replaced by OH via the same mechanism as that with alkenes. The product is an enol, which tautomerizes to the more stable form, which is an aldehyde. Thus, the net result of hydroboration/oxidation of a terminal alkyne is the hydration of an alkyne to an aldehyde with anti-Markovnikov regioselectivity.

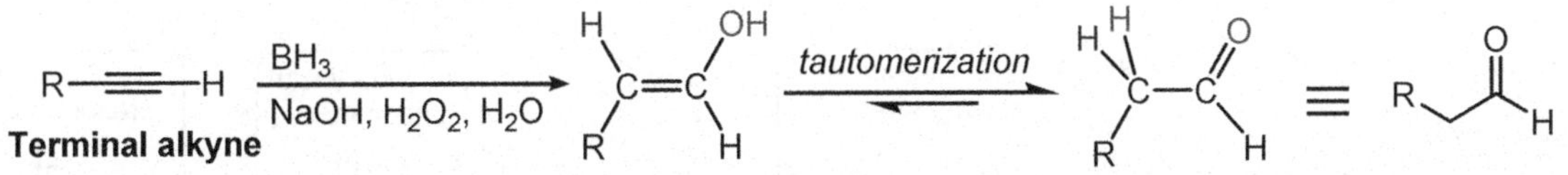

If an internal alkyne (i.e., R–C≡C–R') is used, there is no significant difference in the $\delta+$ on each alkyne carbon, so hydroboration across the C≡C can occur with both orientations. Thus, the net result of the hydroboration/oxidation reaction of this alkyne would also be a mixture of two different ketones: R–C(=O)–CH₂–R' and R–CH₂–C(=O)–R'.

Example 37.2

Draw the major product of the reaction shown below:

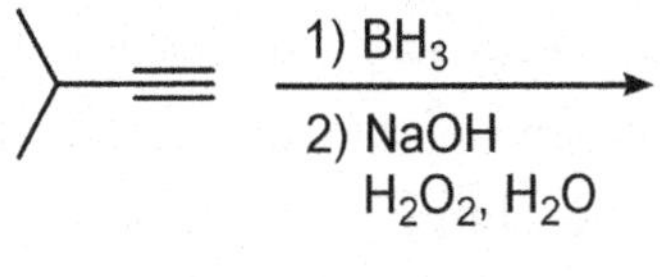

Solution 37.2

We initially form the enol having the OH on the less substituted C. When the enol form tautomerizes to the more stable keto product, an aldehyde is formed on carbon #1:

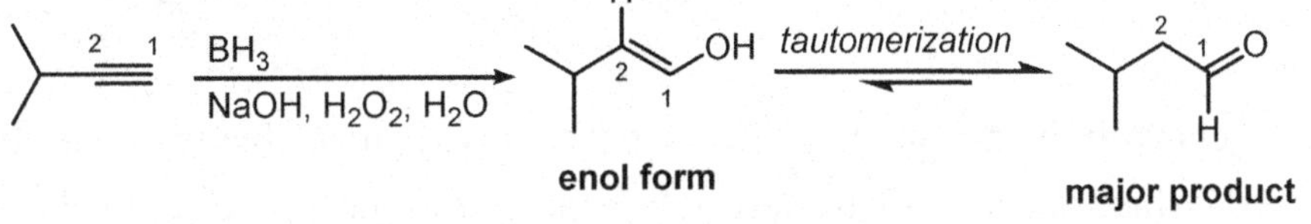

enol form   major product

123

Lesson 38. Radicals and Addition of HBr/ROOR to Alkenes

Lesson 38.1. Definition of a Radical and Radical Stability

A radical is a species having an unpaired electron. As with cations, we will limit the current discussion to carbon-centered radicals. Like carbocations, carbon-centered radicals feature an sp^2-hybridized carbon atom, in which the C does not have an octet, so it will have an affinity for electrons:

General structure of a carbon-centered radical

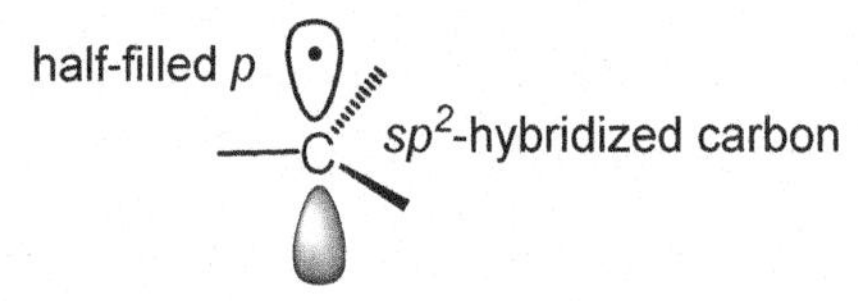

Carbon-centered radicals are similar in electronic structure to carbocations (both lack a complete octet), therefore the trend in radical stability is roughly similar to that for carbocations, but resonance exerts an even higher stabilizing effect in radicals. The stability trends can thus be summarized here:

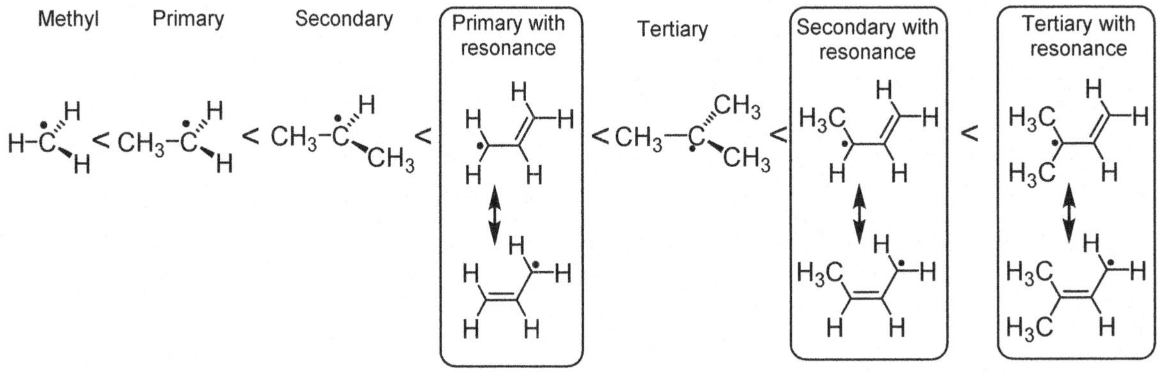

Lesson 38.2. Addition of HBr/RO–OR to Alkenes

When an alkene reacts with HBr in presence of peroxides (RO–OR), the C—C π bonds breaks and two σ bonds form between the carbons of the alkene and the H and Br. The reaction follows anti-Markovnikov rule, where the hydrogen bonds to the more substituted carbon and the Br bonds to the less-substituted carbon. The reaction mechanism involves free-radical intermediates.

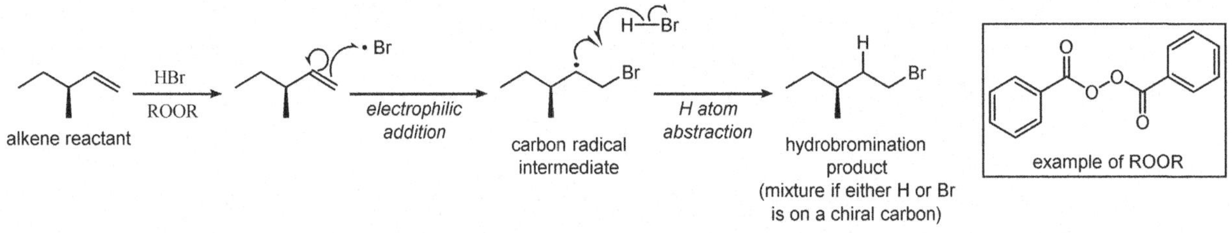

When Br• reacts with the π-bond in an alkene, it will react in such a way that the most stable carbon radical is formed, which occurs when the C–Br bond is formed at the less substituted alkene carbon.

Keep in mind that when an electrophile (i.e., Br$^{\bullet}$) adds to an alkene π-bond, it can do so from above or below the plane. Furthermore, a carbon radical is sp^2-hybridized, and is planar, therefore it can abstract H$^{\bullet}$ from HBr from either above or below the plane. Peroxide-mediated hydrobromination thus is not stereoselective (a mixture of stereoisomers will form).

Lesson 38.3. Mechanism of Addition of HBr/RO–OR to Alkenes

The free-radical mechanism generally involves three main types of steps: initiation, propagation and termination

Initiation steps: The O–O single bond in a peroxide (RO–OR) is very weak and can easily be broken, where its homolysis can often be induced even under fluorescent room lighting or modest heating (e.g., 60 °C). If this homolysis occurs in the presence of HBr, an alkoxy radical (RO$^{\bullet}$) will abstract a hydrogen atom (i.e., H$^{\bullet}$) and generate a bromine radical (Br$^{\bullet}$), which is the radical propagating species. The peroxide most commonly used for these types of reactions is benzoyl peroxide (BPO, shown below).

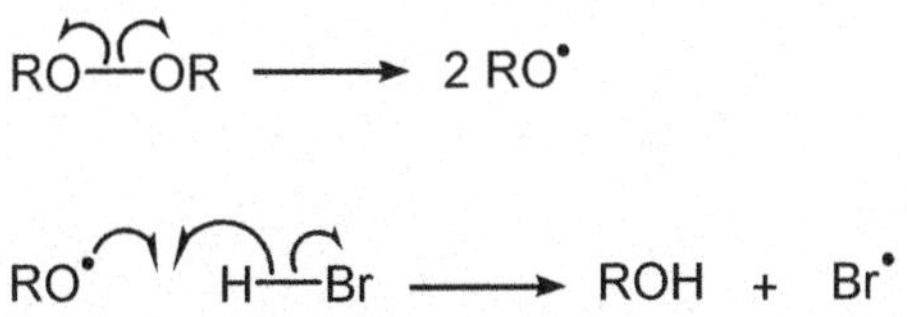

RO—OR $\longrightarrow$ 2 RO$^{\bullet}$

RO$^{\bullet}$ H—Br $\longrightarrow$ ROH + Br$^{\bullet}$

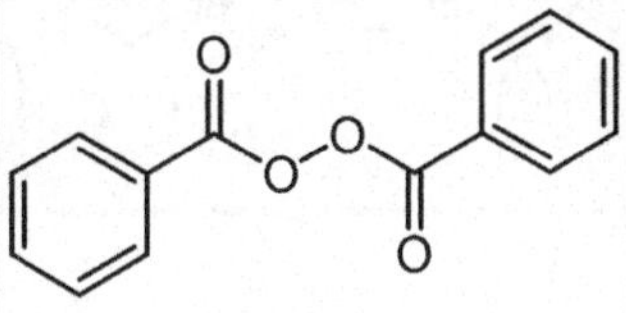

BPO

Propagation steps: As is the case for any atom without a complete octet, Br$^{\bullet}$ will function as an electrophile. If Br$^{\bullet}$ is generated in the presence of an alkene, one of the π-electrons in the C=C bond will move towards Br$^{\bullet}$ and form a C–Br bond, while the other π-electron will stay behind on other carbon and form a carbon radical. The C–Br bond will form at the less substituted alkene carbon, because the carbon radical will be more stable at the more substituted alkene carbon (the trend of carbon radical stability is the same as carbocations). This carbon radical can then react with another molecule of HBr, abstracting H$^{\bullet}$ to form a C–H bond and regenerating the propagating species Br$^{\bullet}$.

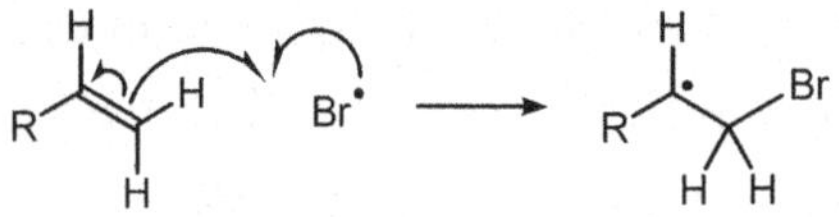

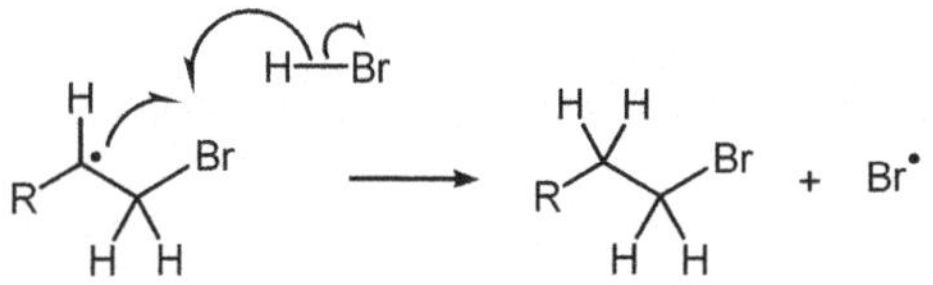

Termination steps: A reaction in which one or more radical species react and afford zero radical products is called a "radical termination reaction". Each radical intermediate can react with another molecule of itself or with one of the other radical intermediates present in the reaction. The possible products are shown below:

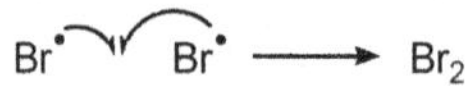

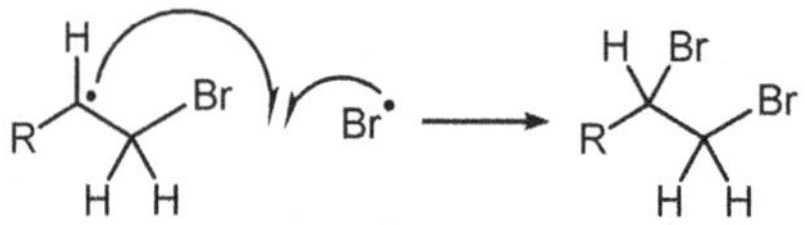

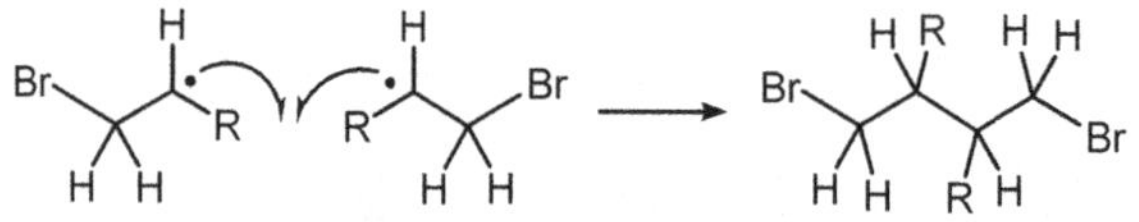

Example 38.1

Draw the major product for the reaction shown below:

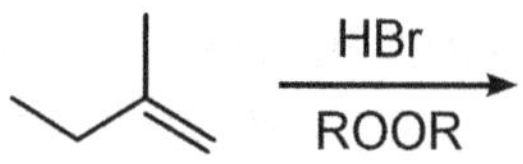

Solution 38.1

Hydrobromic acid, in the presence of ROOR, will generate Br$^{\bullet}$ that does electrophilic addition to the π-bond in 2-methyl-1-butene (step *i*), forming a C–Br bond at the less substituted carbon and a radical at the more substituted carbon. This carbon radical then abstracts an H atom from another molecule of HBr (step *ii*), to form a C–H bond and regenerate the radical propagating species Br$^{\bullet}$. Because the carbon radical is sp^2-hybridized, the C–H bond can be formed either above or below the sp^2 plane, which will afford a 1:1 mixture of enantiomers.

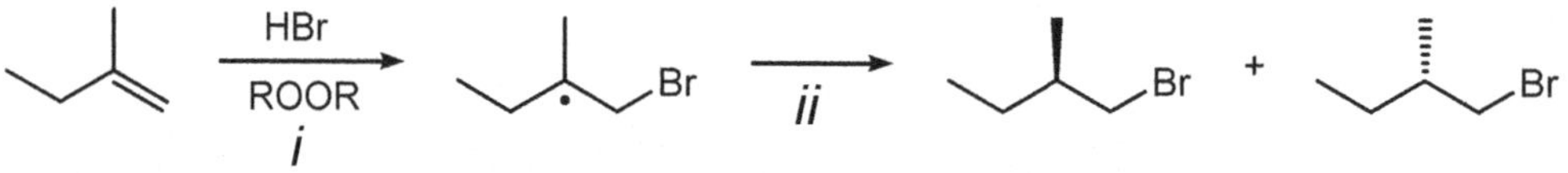

Lesson 39.1. Free Radical Substitution of Alkanes

Alkanes can undergo two reactions, combustion (burning in presence of Oxygen to give carbon dioxide and water) and free radical halogenation in presence of X_2 (where X is mainly Br or Cl) and light or heat to give alkyl halide. The reaction is a substitution reaction where the C–H σ bond on the sp^3 carbon is substituted with a C–X.

Alkane   Alkyl halide

Lesson 39.2 Mechanism of Free Radical Substitution

Initiation step: The X–X bond in X_2 (X = Cl, Br) is relatively weak and can be broken homolytically via photoirradiation (which is typically represented with "$h\nu$"). Homolysis of the X–X bond generates two equivalents of X$^{\cdot}$, which is the radical propagating species.

Propagation steps: When X$^{\cdot}$ encounters an alkane molecule, it will break a C–H bond homolytically (i.e., abstract H$^{\cdot}$) to form a molecule of HX and generate a carbon radical (don't worry about regiochemistry for now). The carbon radical then reacts with a molecule of X_2, breaking the X–X bond homolytically to form a C–X bond and regenerate the radical propagating species X$^{\cdot}$. Because the concentration of X$^{\cdot}$ is usually low, these reaction conditions will typically only afford the monohalogenated product.

Termination steps: The radical termination reactions in the free radical halogenation of alkanes are shown below:

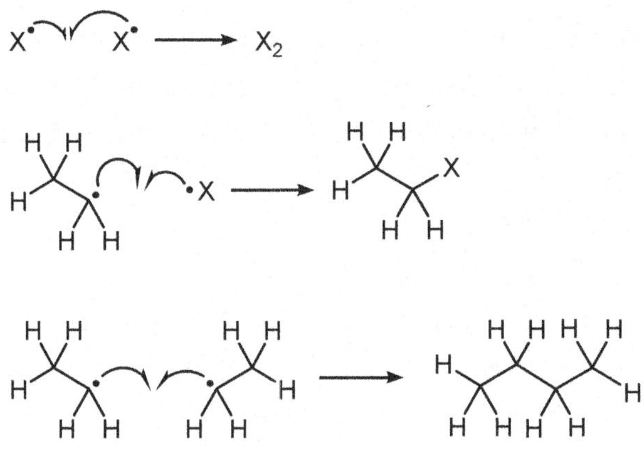

Example 39.1

A carbon radical can also undergo an elimination reaction that terminates radical propagation. Draw one such termination reaction for the radical bromination of 2-methylpropane.

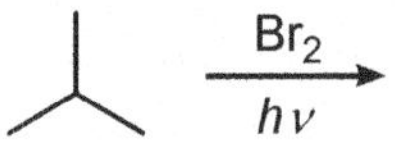

Solution 39.1

The Br• generated from photoirradiation of Br_2 will abstract H• from 2-methylpropane (step *i*), leaving behind a radical on the 3° carbon. Homolysis of a C–H σ-bond adjacent to the radical (step *ii*) yields an electron that can pair with the carbon radical electron to form a C=C π-bond. This elimination reaction produces an alkene and H•, which dimerizes to H_2.

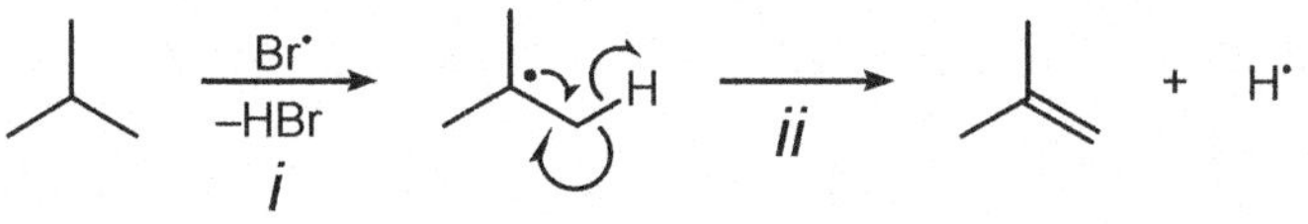

Lesson 39.2 Regiochemistry and Stereochemistry

Remember that the radical halogenation of an alkane will only replace C–H bonds with C–X bonds, it will never break C–C bonds. So, the easiest way to draw all the possible products of this reaction is to identify all of the chemically unique C–H bonds in a given molecule. If our reactant is 2-methylbutane, there are 4 chemically unique C–H bonds, thus radical halogenation of this molecule will generate 4 different monohalogenated products (which are constitutional isomers of each other). To identify which will be the major product, we need to consider the relative stability of the radicals which led to each constitutional isomer. Abstraction of H• from position (1) or (4) will yield a 1° radical, from position (2) will yield a 3° radical, and from position (3) will yield a 2° radical. Given that the stability

128

of carbon radicals is 3° > 2° > 1°, intermediate (2) is the most stable, and thus we would expect the major product to be derived from that intermediate. When X = Br, product (2) is by far the most abundant. However, when X = Cl, products (1) and (4) are formed in nearly equal amount to product (2)! So, it is not only intermediate stability that is influencing product distribution.

Going back to our example with 2-methylbutane, we have only considered the constitutional isomers, but we can see that 2-bromo-3-methylbutane contains a chiral center. If we consider carbon radical (3), which gave rise to this product, we can see that the radical resides on an sp^2-hybridized carbon, with the "missing" electron in an unhybridized p-orbital. When this radical reacts with a molecule of Br_2, it can form a C–Br bond from either above or below the sp^2 plane. Because there is no difference in energy between the two approaches, we would get a 1:1 mixture of enantiomers. Thus, the net reaction is the conversion of an alkane and X_2 to alkyl halide and HX via a mechanism that is neither regioselective nor stereoselective.

Example 39.2

Draw the monohalogenated products for the reaction shown below. Indicate which is major

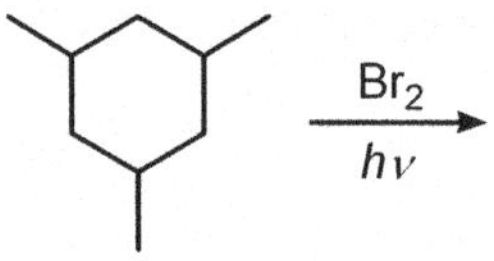

Solution 39.2

In this reactant we have three different hydrogens that will give three different alkyl halides upon free radical halogenation. The major product will be the tertiary alkyl halides rather than the secondary or primary one.

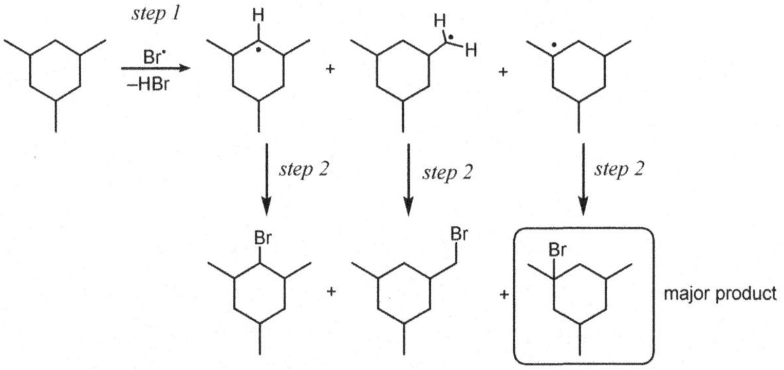

Lesson 40. Aromaticity: A Highly Stabilizing Effect

Lesson 40.1. Aromaticity is a Special Type of Resonance Stabilization

In Lesson 5, we saw that resonance delocalization provides additional stabilization to a molecule. If we have a molecule with alternating single and π-bonds, that molecule will be able to engage in resonance by moving the π-bonding electrons around (or *delocalizing* them) through resonance. A molecule with alternating single and π-bonds is called a π-conjugated molecule. In this Lesson, we study a special case of stabilization of a π-conjugated system called **aromaticity**. This special type of resonance delocalization that can only occur in a cyclic π system. A molecule that exhibits aromaticity is said to be **aromatic**. The canonical aromatic compound is benzene (C_6H_6):

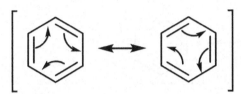

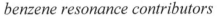

benzene resonance contributors   *resonance hybrid*   *another way to represent benzene*

As illustrated in the graphic above, benzene has two resonance contributors that are identical in energy. The resonance hybrid more accurately depicts the facts that (1) the electrons are delocalized evenly around the ring and (2) rather than alternating C–C single and double bonds, all 6 C–C bonds in benzene instead have bond orders of 1.5. To emphasize the uniformity of the electron density and bonding throughout the ring, benzene is sometimes drawn with a circle in the middle of the ring. Despite having partial C=C character, the **C–C bonds in aromatic systems are so much more stable than in alkenes that they will not undergo typical alkene reactions**. In terms of physical properties, the boiling point of benzene is similar to that of other six-membered ring hydrocarbons:

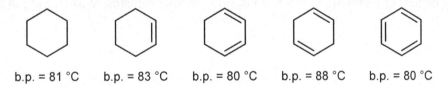

b.p. = 81 °C   b.p. = 83 °C   b.p. = 80 °C   b.p. = 88 °C   b.p. = 80 °C

In contrast, larger-ring aromatic compounds comprised by more than one benzene ring fused together tend to be solids and to have higher melting points than their saturated analogues:

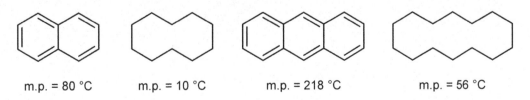

m.p. = 80 °C   m.p. = 10 °C   m.p. = 218 °C   m.p. = 56 °C

131

Example 40.1

Provide the product of this reaction:

Solution 40.1

These conditions are used for the hydrogenation of alkenes. Hydrogenating the C=C bonds in the benzene would disrupt aromaticity, which is too energetically costly, so only the non-aromatic C=C bond undergoes hydrogenation to afford the product:

Lesson 40.2. Criteria for Aromaticity

Benzene is not the only molecule that exhibits the special type of stability that is called aromaticity. An aromatic molecule may have atoms other than C in it or may even be ionic. There are certain criteria for a molecule to exhibit aromaticity:

1. The molecule must be cyclic and planar
2. The π-electrons must be delocalized around the whole ring. This requires each atom to have either a lone pair, a π-bond, or an empty p-orbital
3. The number of electrons that are delocalized around the ring must be $4n+2$ (where n is an integer). This last criterion is called **Hückel's rule**.

A molecule that meets criteria 1 and 2 for aromaticity, but instead has $4n$ electrons, is **antiaromatic**. A molecule that is neither aromatic nor antiaromatic is simply referred to as **nonaromatic**.

Example 40.2

Identify each of the following as being aromatic (A), antiaromatic (AA) or nonaromatic (NA):

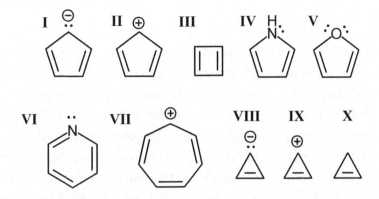

Solution 40.2

We must examine each molecule to see what classification is correct. They are all cyclic and, although there are some monocyclic π-conjugated molecules that are not planar (ring size 8 or greater where the number of π electrons $4n+2$), these are also planar. Let us see whether each of these molecules has resonance delocalization around the ring and then check for how many electrons are delocalized:

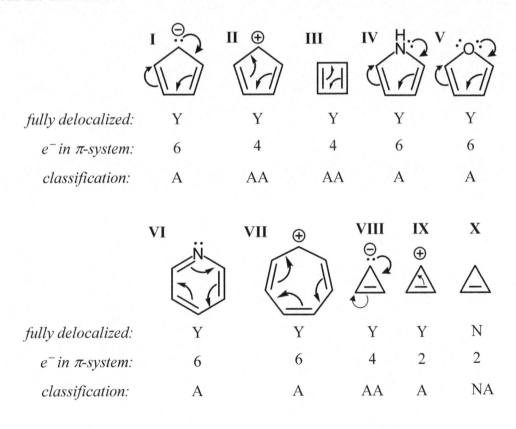

	I	II	III	IV	V
fully delocalized:	Y	Y	Y	Y	Y
e^- *in π-system:*	6	4	4	6	6
classification:	A	AA	AA	A	A

	VI	VII	VIII	IX	X
fully delocalized:	Y	Y	Y	Y	N
e^- *in π-system:*	6	6	4	2	2
classification:	A	A	AA	A	NA

A few points are worth noting. First, delocalization around the entire ring can only occur if each carbon has an unhybridized p-orbital (empty or in a π bond) or a lone pair (remember that resonance structures involve the movement of π-bonds and lone pairs). Each molecule except **X** satisfies this criterion.

Another important point has to do with the lone pairs on the O and N atoms of IV, V and IV. In VI, the lone pair on N does not participate in the resonance delocalization. This is because the N in VI already has a π bond to it; it does not need to use its lone pair to participate in resonance. In fact, the lone pair on the N in VI is in an sp^2-hybridized orbital at 90° angle to the π system. We only count the electrons in the π system (the number delocalized), so the lone pair electrons are not counted in our check for Hückel's rule. In contrast, the N in IV does not have a π bond as it is drawn, so its lone pair is in an orbital that can participate in the π system, so these electrons are counted. In V, notice that there are two lone pairs on the O atom. Only one lone pair can be delocalized into the π-system, whereas the other is at 90° to the π-system, so only one lone pairs is counted in our assessment of the number of electrons in the π system.

Lesson 41. Aromaticity Effects on Acidity and Basicity

41.1. Aromaticity Effects on Acidity

As we saw in Lesson 40, some anions are aromatic. Aromatic anions are significantly more stable than anions that are not aromatic (e.g., H_3C^-). In Lesson 7, we learned that more stabilized anions are weaker bases, and consequently they are the conjugate bases derived from stronger acids. With these facts in mind, it is unsurprising that hydrocarbons which yield aromatic anions upon deprotonation are significantly stronger acids than hydrocarbons which do not. Consider the following pK_a values:

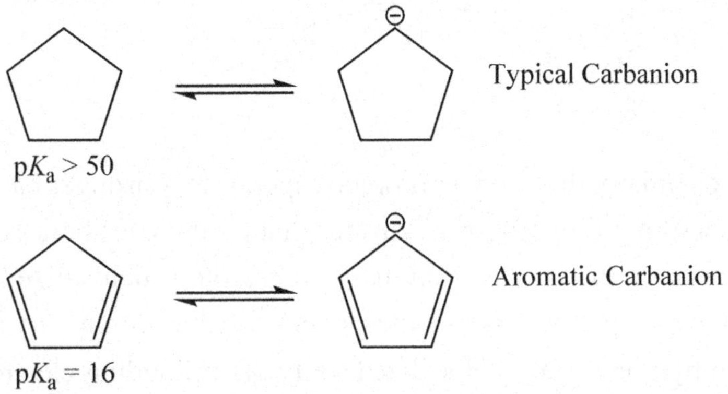

As you can see, cyclopentadiene has a much lower pK_a than does cyclopentane. The remarkable stability of its conjugate base, as a result of its aromatic nature, makes cyclopentadiene over 10^{34} times more acidic than cyclopentane! We must be careful to consider whether an anion is aromatic or not when we are asked a question that requires us to consider the thermodynamic favorability of a reaction.

41.2. Aromaticity Effects on Basicity

Recall from Lesson 6 that a Lewis base is an electron pair donor. The ability of a compound to donate its lone pair and act as a Lewis base can be influenced significantly by whether or not a lone pair is tied up in resonance or in an aromatic system. In cases where the lone pair is involved in aromaticity, pulling that lone pair away to do a Lewis acid–base reaction would thus remove aromatic stabilization. We know that aromaticity provides a large amount of stabilization, so such an acid–base reaction would be thermodynamically unfavorable. Consider the basicity of pyridine versus that of pyrrole:

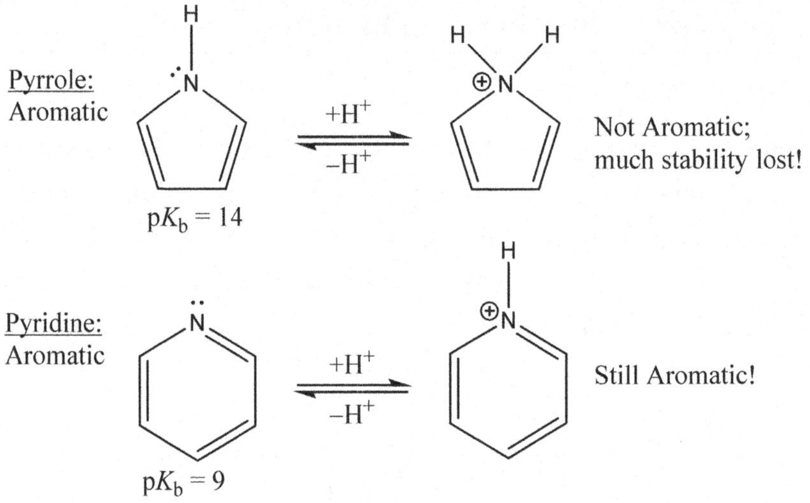

Pyrrole: Aromatic — $pK_b = 14$ — +H⁺ / −H⁺ — Not Aromatic; much stability lost!

Pyridine: Aromatic — $pK_b = 9$ — +H⁺ / −H⁺ — Still Aromatic!

Recall from general chemistry that a lower pK_b corresponds to a stronger base. The higher pK_b value of pyrrole (14) versus that of pyridine (9) thus indicates that pyridine is about 100,000 times more basic than is pyrrole. This is because the lone pair on the N in pyrrole is delocalized as a key component of the aromatic system. If the lone pair instead donates to an acid, this aromaticity is disrupted. In contrast, the lone pair on the N in pyridine is not delocalized (only the π-bonding electrons are delocalized), so the product of protonation is still aromatic, and thus protonation of pyridine is a much more favorable reaction than is protonation of pyrrole. The general concept to take away from this Lesson is that **reactions that create aromaticity in products tend to be thermodynamically favorable, whereas reactions that disrupt the aromaticity of a starting material tend to be thermodynamically unfavorable.**

Example 41.1

A) Which is the most basic nitrogen atom in this molecule?

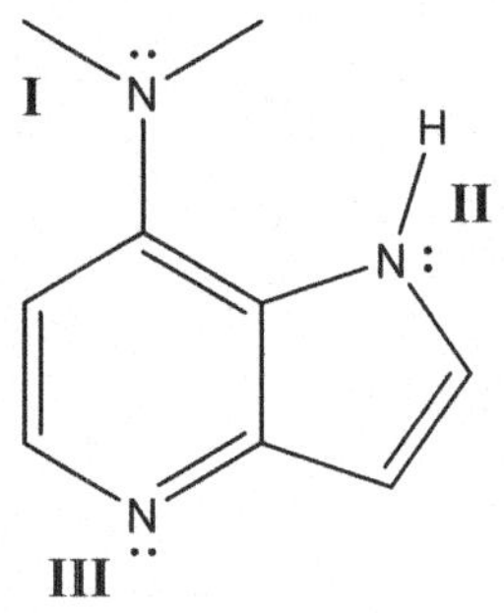

B) Which will undergo heterolysis of the C–Br bond (to form a carbocation and bromide) most rapidly?

136

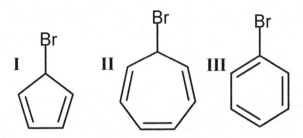

I   II   III

Solution 41.1

A) We must consider whether each lone pair participates in resonance delocalization or not. We first check the lone pair on nitrogen **I** and see that it can participate in resonance:

The lone pair on nitrogen **I** is engaged in resonance delocalization. To take that lone pair out of the delocalized system and donate it to an acid instead would cost the resonance delocalization energy. This will not be a very basic site. Next, we check nitrogen **II**:

Not only is the lone pair on nitrogen **II** engaged in delocalization, but it is required to make the molecule aromatic. To remove this lone pair would require significant energy to break up the aromaticity. This is an even less basic nitrogen than is nitrogen **I**. The lone pair on nitrogen **III**, however, is not needed for resonance delocalization or for aromaticity. Nitrogen **III** will be by far the most basic N in the molecule.

137

The carbocations produced from C–Br heterolysis in **I–III** are:

I $\xrightarrow{-Br^-}$ **IV** *Antiaromatic! Very unstable!*

II $\xrightarrow{-Br^-}$ **V** *Aromatic carbocation! Gains significant stability; fastest heterolysis reaction.*

III $\xrightarrow{-Br^-}$ **VI** *Carbocation that is not stabilized by resonance!*

Only heterolysis of **II** leads to the formation of a highly-stabilized aromatic carbocation (**V**), so this is the fastest reaction.

Lesson 42.1. General Rules for Monosubstituted Benzene

For many common substituents, the rules for naming a monosubstituted benzene are the same as for a monosubstituted cyclohexane, but with benzene as the parent molecule instead of cyclohexane. Consider these examples:

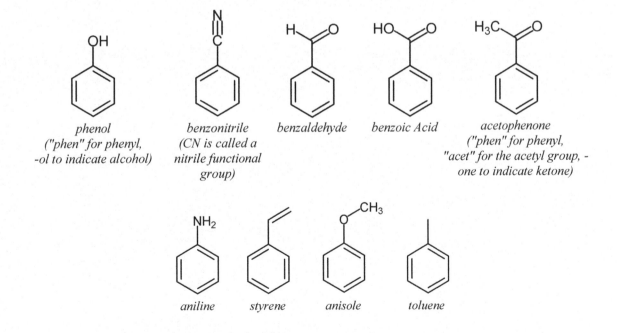

fluorobenzene	*chlorobenzene*	*ethylbenzene*	*nitrobenzene*	*isobutylbenzene or (2-methylpropyl)benzene*

Note the new functional group, NO_2, termed a nitro group. In cases where a **benzene ring is itself a substituent of a larger molecule, it is called a "phenyl" group** and is given the abbreviation "Ph". The –CH_2Ph group is called "benzyl" and is abbreviated Bn.

Lesson 42.2. Common Names for Some Monosubstituted Benzene Derivatives

Some monosubstituted benzene compounds have specific common names that do not conform to systematic nomenclature rules. Many of these names have been used for centuries and are simpler than the systematic names, so they are still widely-used today. Because these non-systematic names are still used by chemists, these names must be memorized. The non-systematic names for monosubstituted benzene compounds that you will be expected to know are shown below:

phenol
("phen" for phenyl,
-ol to indicate alcohol)

benzonitrile
(CN is called a
nitrile functional
group)

benzaldehyde

benzoic Acid

acetophenone
("phen" for phenyl,
"acet" for the acetyl group, -
one to indicate ketone)

aniline   *styrene*   *anisole*   *toluene*

These are the most frequently encountered common names for monosubstituted benzenes that you are likely to encounter in introductory organic chemistry course. For some molecules, the non-systematic name is a logical reflection of the molecular structure (e.g., phenol is an alcohol on a phenyl group), whereas for others the name provides no such information.

Example 42.1

Provide IUPAC names for each of the molecules shown below:

I

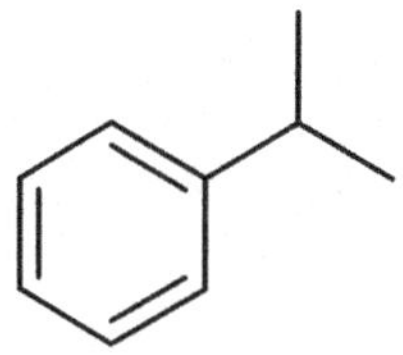

II

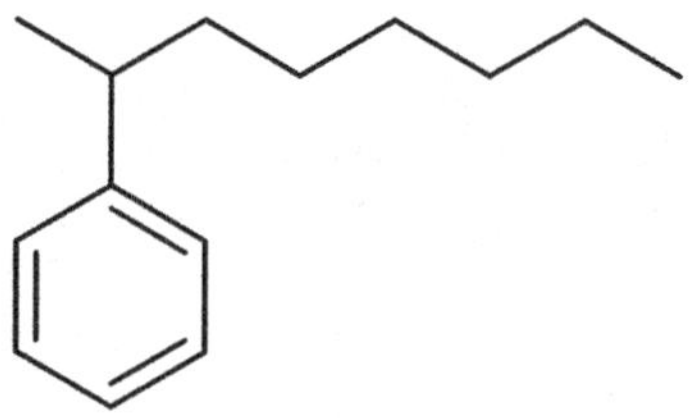

Solution 42.1

For molecule **I**, the parent is "benzene" and it has one substituent, which is an "isopropyl" group, so the molecule is properly named isopropylbenzene.

For molecule **II**, the octane chain has more carbon atoms than the benzene ring, so we must use "octane" as the parent. Recall that when benzene is a substituent on a larger molecule, it is called a "phenyl" group. So, molecule **II** is properly named 2-phenyloctane.

43.1. Systematic Nomenclature

Polysubstituted benzene derivatives can be named using the same approach as for polysubstituted cyclohexane derivatives, but without the need for any *cis-* or *trans-* labels, because benzene is planar. Here are some examples:

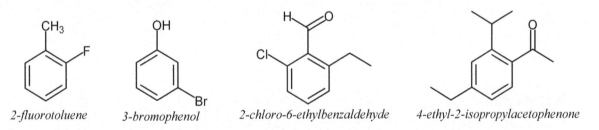

1,2-difluorobenzene   1-bromo-2-chlorobenzene   1-chloro-3-ethyl-2-isopropylbenzene   2-ethyl-4-iodo-1-propylbenzene

43.2. Special Parent Structures

We saw in the previous Lesson that some monosubstituted benzene derivatives have historical common names that are now deeply rooted in the field of chemistry. When these structures are present in a polysubstituted benzene derivative, one must use the common name as the parent rather than using benzene as the basis for the name. In the common name parent structure, the substituent from which its common name arises is always given the number 1. In phenol, for example, the C bearing the OH group is assigned the number 1, in toluene, the C with the methyl group is assigned the number 1, etc. Here are a few examples to illustrate:

2-fluorotoluene   3-bromophenol   2-chloro-6-ethylbenzaldehyde   4-ethyl-2-isopropylacetophenone

43.3. Alternative Nomenclature for Disubstituted Benzene Derivatives

For disubstituted benzene derivatives, one can use the systematic nomenclature as illustrated above. An alternative nomenclature system can also be used *for disubstituted benzene derivatives only*. Instead of using 1,2-, one can simply use *ortho-*. In the case of an *ortho-*disubstituted benzene derivative, the *o-* label is placed in front of the rest of the name, and the numbers are omitted. A similar convention allows us to use *m-* (for *meta-*) in place of numbers for 1,3- derivative or *p-* (for *para-*) in place of numbers for 1,4- derivatives.

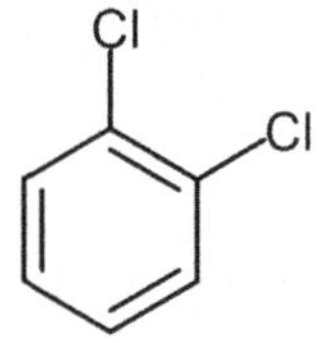

o-dichlorobenzene 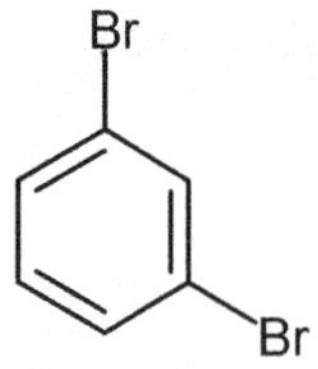   m-dibrombenzene 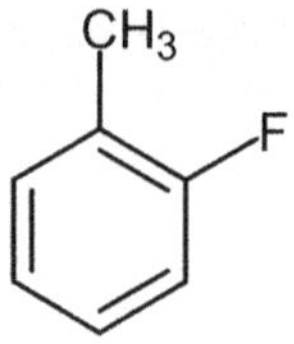   o-fluorotoluene 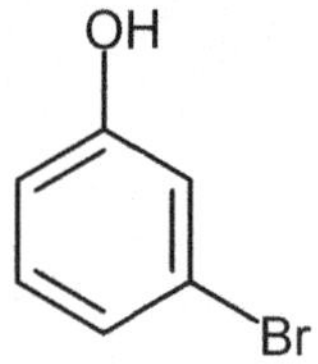   m-bromophenol 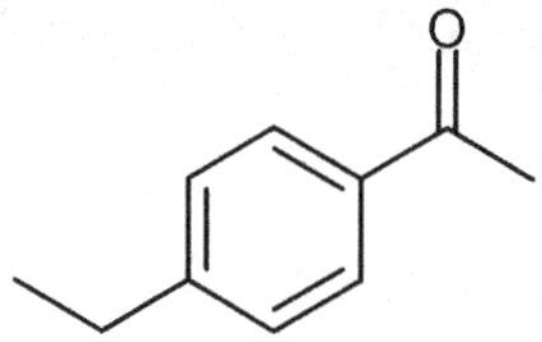   p-ethylacetophenone

As for monosubstituted benzene derivatives, there are some disubstituted benzene derivatives that have long-standing common names that must be committed to memory. The most frequently-encountered examples in organic chemistry are the "xylenes", which are "dimethylbenzene" derivatives:

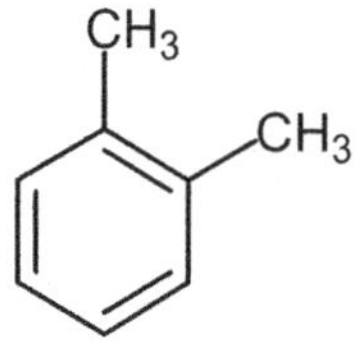

o-xylene 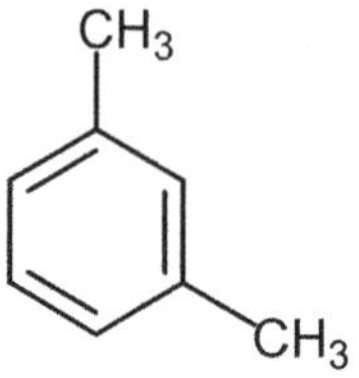   m-xylene 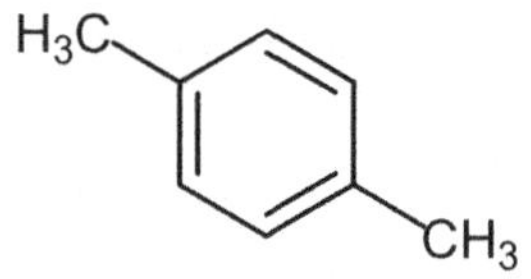   p-xylene

---

Example 43.1

Provide the proper names for each of these molecules:

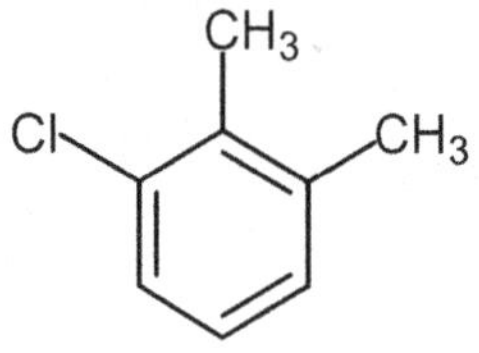

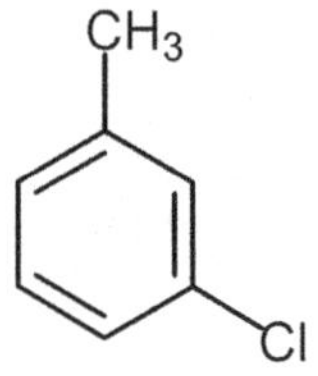

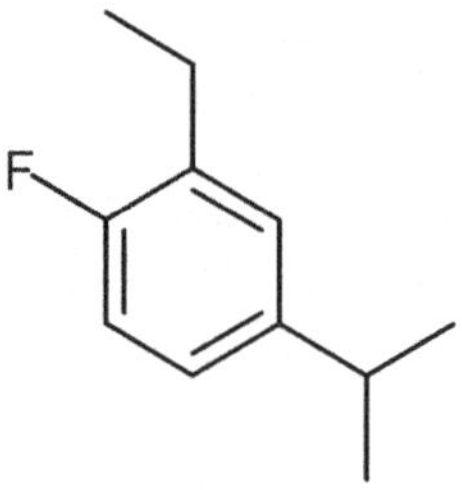

Solution 43.1

One challenge is to identify whether the parent has a non-systematic common name, such as xylene (leftmost molecule) or toluene (center molecule). The proper names should be:

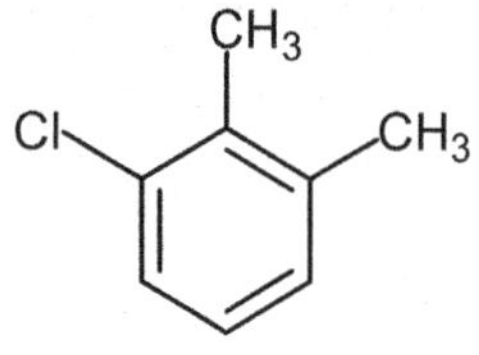

3-chloro-o-xylene 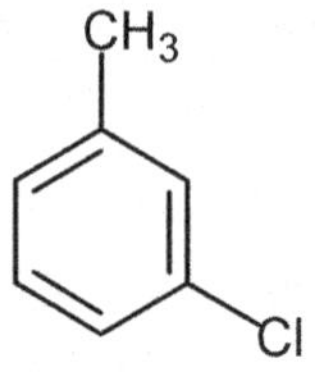   m-chlorotoluene
or
3-chlorotoluene   2-ethyl-1-fluoro-4-isopropylbenzene

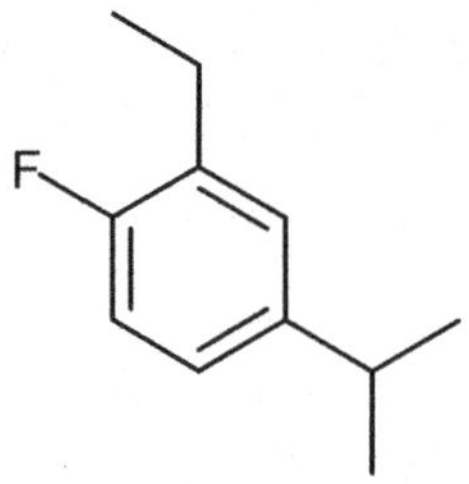

44.1. Electrophilic Aromatic Substitution (EAS)

The C=C bonds in benzene are much less reactive than are isolated alkenes due to the extra stability that aromaticity affords (recall that stability and reactivity are inversely related). Because the π-bonds in benzene are less reactive (i.e., less nucleophilic) than those in non-conjugated alkenes, a more reactive electrophile is required to draw the π-bonding electrons out of a benzene ring to initiate a reaction. One type of reaction between an aromatic system and an electrophile is **electrophilic aromatic substitution** (**EAS**). As its name suggests, this reaction requires an electrophile and results in the substitution of an H on the benzene ring with the electrophile. EAS consists of two steps: 1) electrophilic addition of an electrophile to one of the π-bonds followed by 2) electrophilic elimination of H^+ from the C to which the electrophile added:

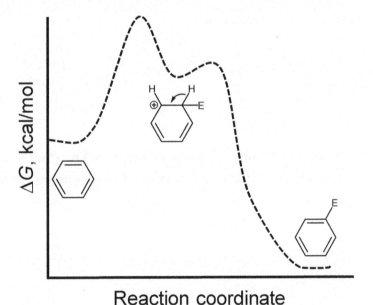

A general reaction coordinate diagram for this reaction would look like this:

We will cover several specific types of EAS in this Lesson, but they all have the same underlying mechanism. These EAS reactions differ only in the identify of electrophile that adds to the aromatic system in the first step. In the next several Lessons, we will thus look at seven specific sets of reagents and see how they generate electrophiles that are sufficiently reactive to add to the π-bonds in benzene.

44.2. Carbocations as Electrophiles in EAS – Friedel-Crafts Alkylation.

The first class of electrophile we will consider that undergoes electrophilic addition to benzene is a carbocation. When a carbocation is the electrophile for EAS, the reaction is called **Friedel–Crafts alkylation**:

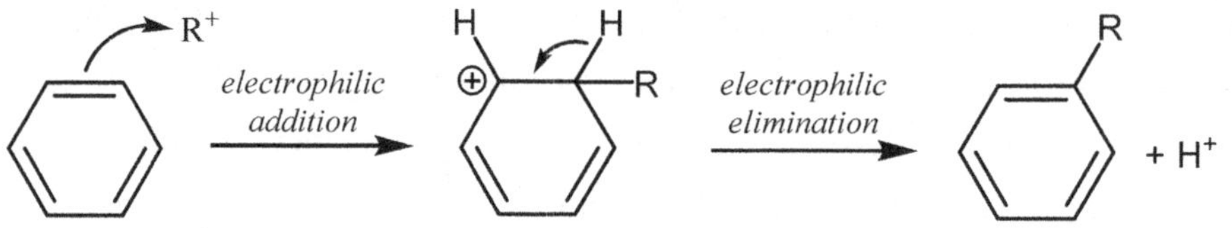

We will focus on a couple ways to generate carbocations for the Friedel–Crafts alkylation: 1) electrophilic addition of H^+ to the π-bond in an alkene (Lesson 30) and 2) abstraction of Cl^- from an alkyl chloride by $AlCl_3$. Here is a recap of these ways that one can generate carbocations for use in Friedel–Crafts alkylation:

1) Alkene in the presence of acid

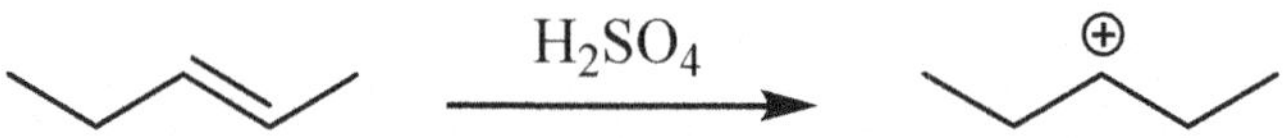

2) Alkyl chloride in the presence of $AlCl_3$

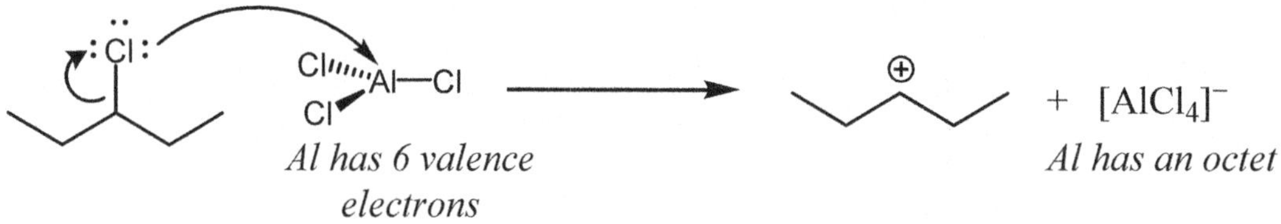

Chloride abstraction by $AlCl_3$ is the **most commonly-employed method to generate a carbocation** for Friedel–Crafts alkylation, and is so effective that **even primary carbocations can be generated in this way**. The various ways that the carbocation-generating reactions shown above may be harnessed for Friedel–Crafts alkylation reactions are shown below:

1) Carbocation is from an Alkene in the presence of Acid

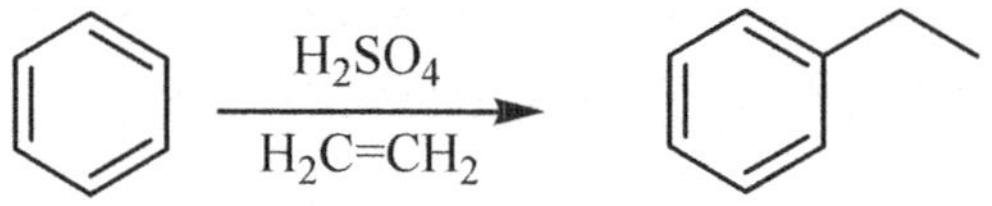

2) Carbocation is from Alkyl chloride in the presence of $AlCl_3$

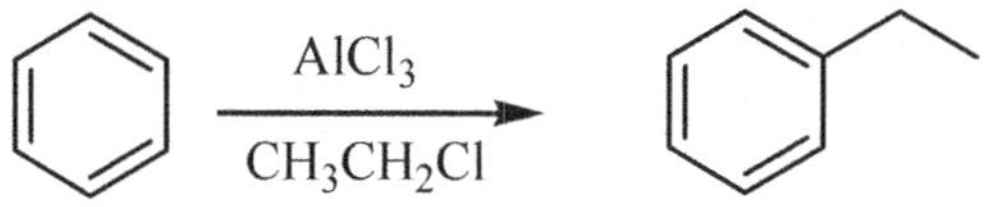

Reaction 2 is the most-used EAS reaction in the world today, because the ethylbenzene product is a key intermediate in the synthesis of polystyrene (the main component of Styrofoam and other common plastics).

Example 44.1

Provide the major organic product for each of these reactions:

A)

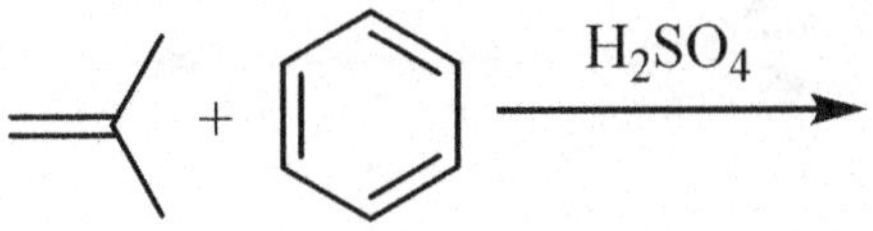

$$H_2SO_4$$

B)

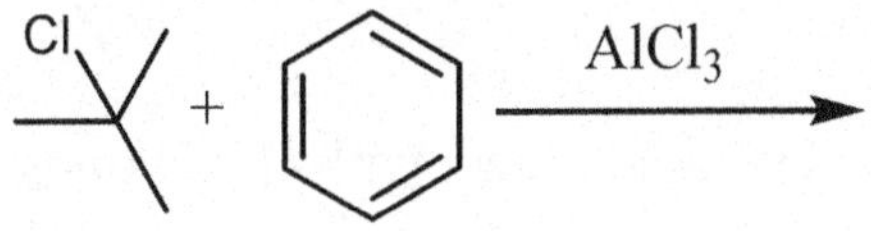

$$AlCl_3$$

Solution 44.1

A) We see that the alkene will produce a cation upon electrophilic addition of a proton from the acid. The more substituted carbocation is formed:

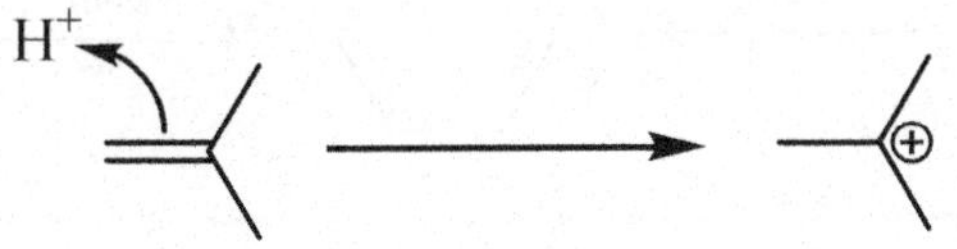

We use the carbocation as the electrophile for electrophilic aromatic substitution with the benzene:

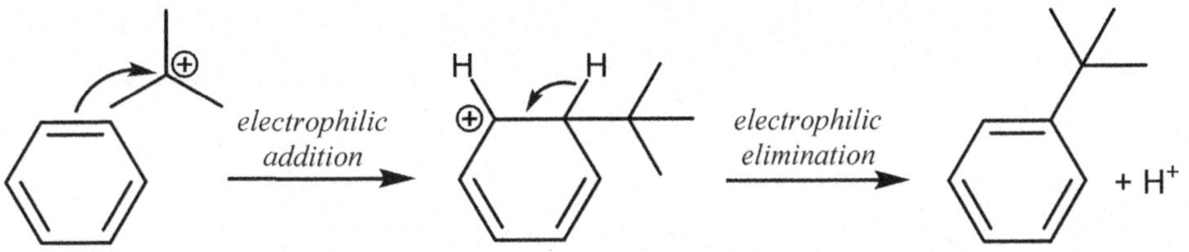

electrophilic addition   electrophilic elimination   + H⁺

B) The AlCl₃ will abstract a chloride from the alkyl chloride to generate a carbocation:

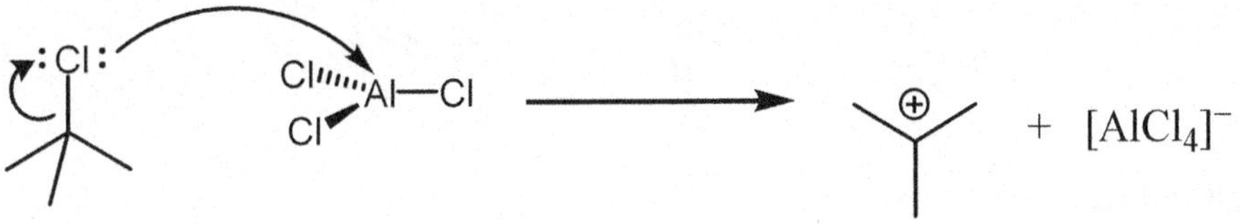

+ [AlCl₄]⁻

145

This carbocation – the same cation we used for EAS in part A – will then participate in the EAS mechanism as shown in the solution to part A.

44.3. Acylium Cations as Electrophiles in EAS – Friedel-Crafts Acylation.

Chloride abstraction by $AlCl_3$ also works with acid chlorides to generate acylium ions:

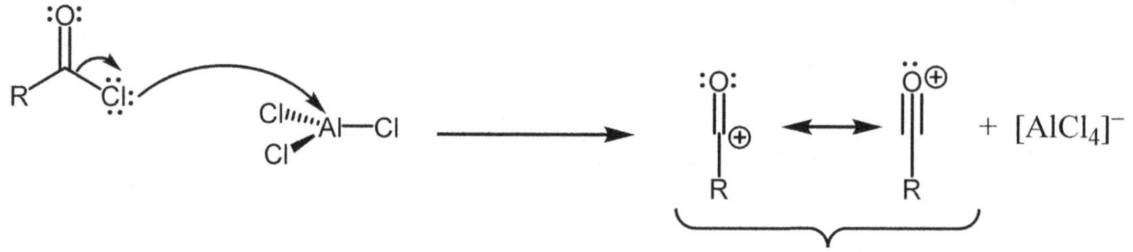

Acylium ion
(two resonance contributors)

Acylium ions are reactive enough to react with C=C bonds in benzene. When an acylium ion is used as the electrophile in an EAS reaction, the reaction is called **Friedel-Crafts acylation**:

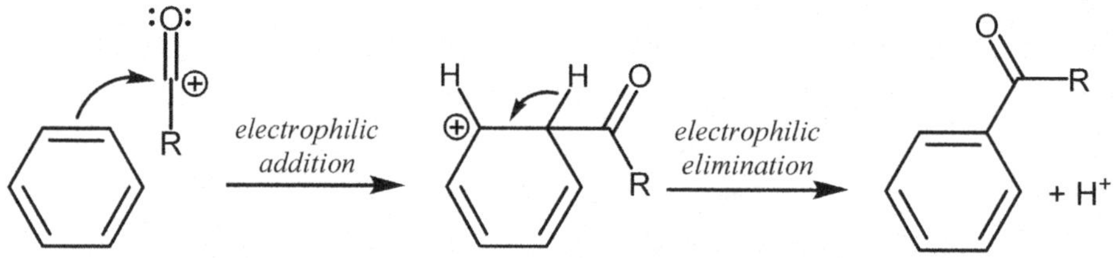

Example 44.2

Draw the major organic product for each of the following reactions:

A)

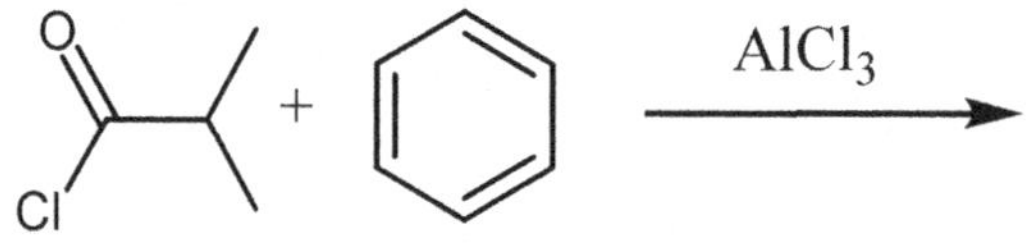

B)

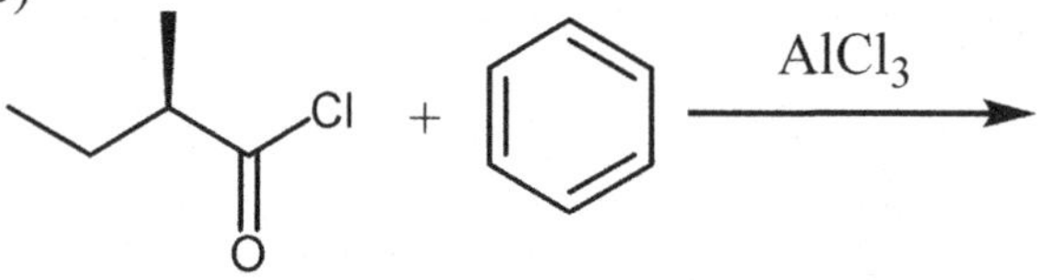

Solution 44.2

A) The $AlCl_3$ will abstract a Cl^- from the acid chloride to form an acylium ion:

146

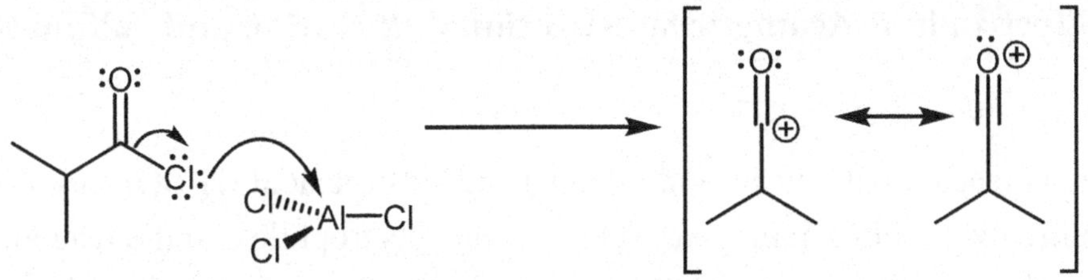

The acylium ion then acts as the electrophile for an EAS reaction:

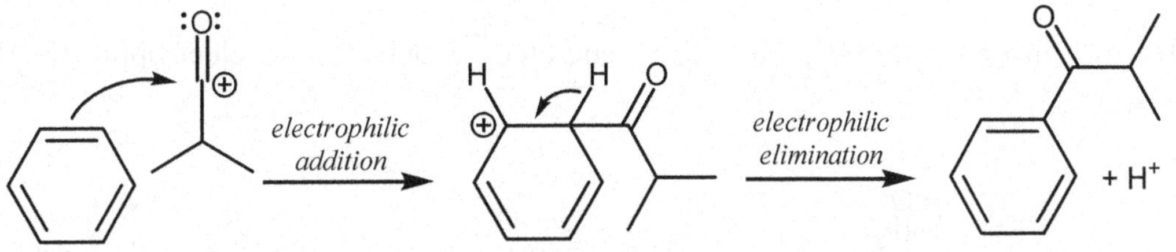

B) The first step is again generation of the acylium cation:

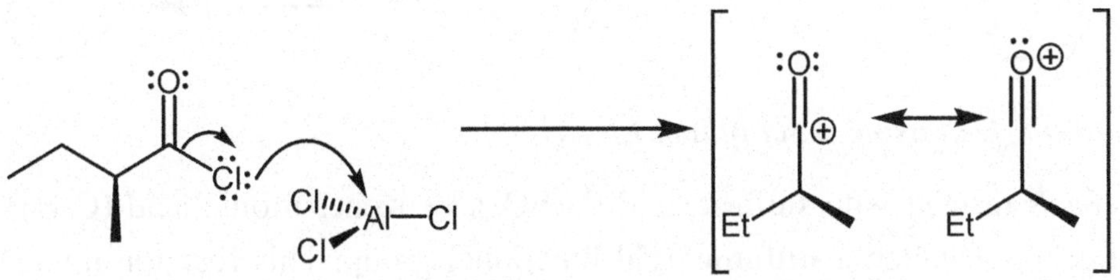

Note that the chiral center in the acid chloride is retained in the acylium cation, because that carbon remains sp^3-hybridized throughout the reaction. It will maintain its configuration even after undergoing EAS:

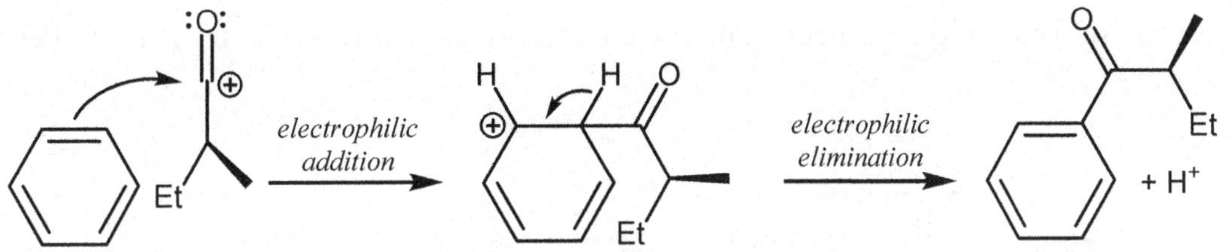

45.1. Nitration: EAS using a Nitronium Electrophile

If benzene is reacted with nitric acid (HNO_3) and sulfuric acid (H_2SO_4), nitrobenzene is produced. This reaction involves a nitronium ion ($[NO_2]^+$) as the electrophile and the reaction is called **nitration**. The nitronium electrophile is generated by the protonation of nitric acid by sulfuric acid as follows:

$$H_2SO_4 + HNO_3 \rightarrow [NO_2]^+ + H_2O + HSO_4^-$$

After the nitronium (NO_2^+) has been generated, it acts as the electrophile in the usual EAS mechanism:

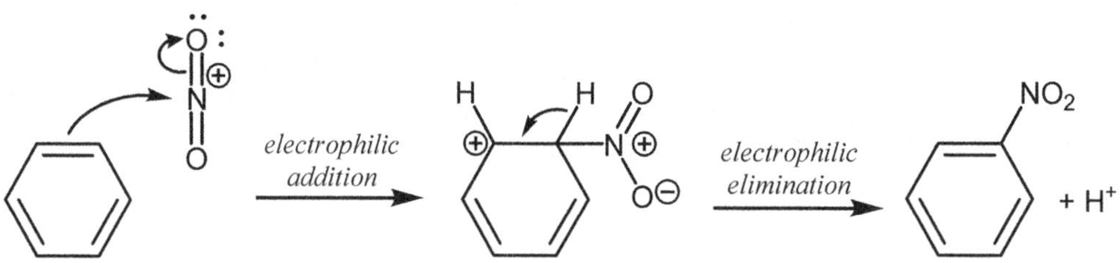

45.2. Sulfonation: EAS using a Sulfonium Electrophile

If benzene is reacted with sulfuric acid (H_2SO_4), benzenesulfonic acid ($C_6H_6SO_3H$) is produced. The –SO_3H group is called a **sulfonic acid** functional group. This reaction involves a sulfonium ion ($[HSO_3]^+$) as the electrophile, and the reaction is called **sulfonation**. The sulfonium electrophile is generated by reaction of sulfuric acid as follows:

$$H_2SO_4 + H_2SO_4 \rightarrow [SO_3H]^+ + H_2O + HSO_4^-$$

After the sulfonium ion has been generated, it acts as the electrophile in the usual EAS mechanism:

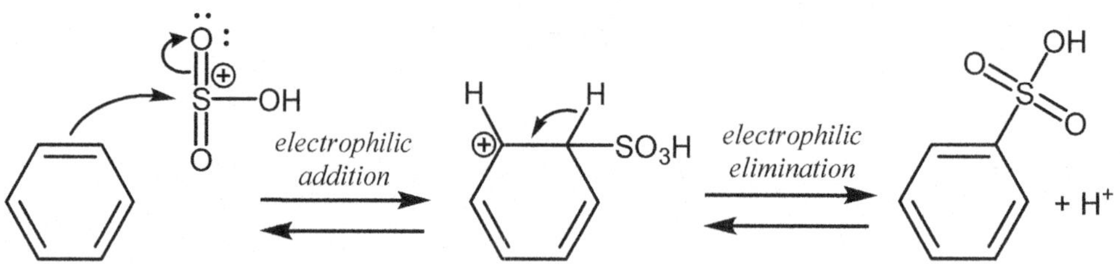

Unlike the other EAS reactions we have seen, sulfonation is a highly reversible reaction, so if an aromatic compound with a sulfonic acid group is heated in the presence of water with trace acid, the sulfonic acid group is removed and replaced with a proton.:

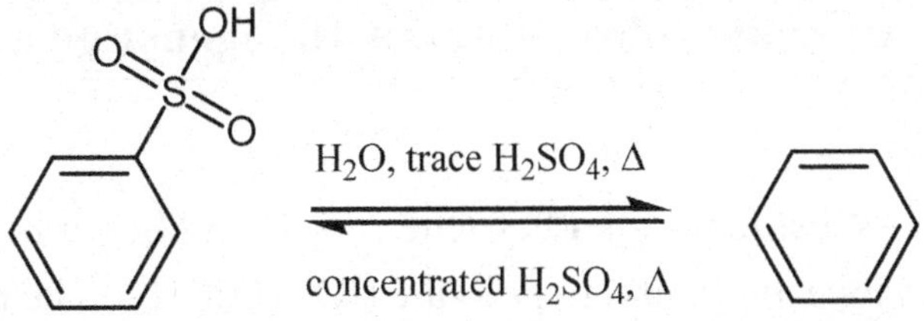

H_2O, trace H_2SO_4, Δ

concentrated H_2SO_4, Δ

Example 45.1

Draw the major organic product for each of these reactions:

A) H_2SO_4, Δ

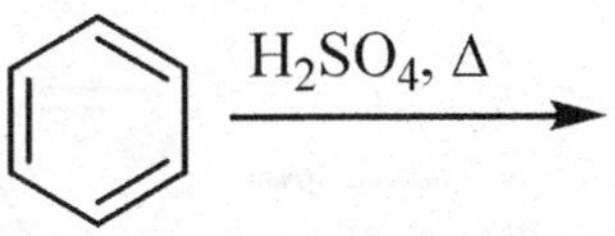

B) H_2SO_4, HNO_3

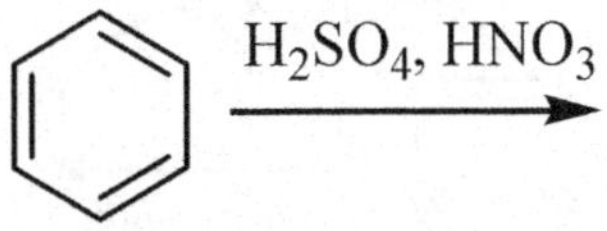

C) dil. H_2SO_4

H_2O

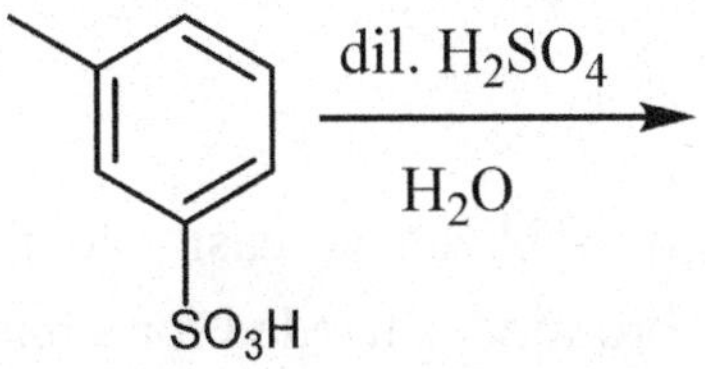

Solution 45.1

A) This is a sulfonation reaction. The product is:

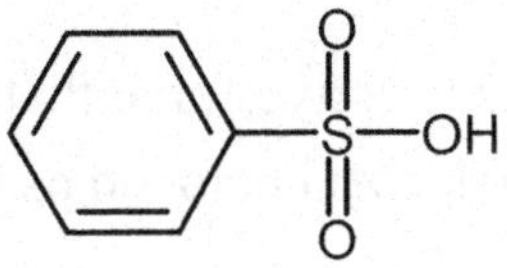

B) This is a nitration. The product is:

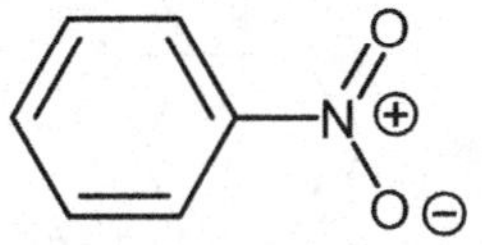

C) This is the reverse of sulfonation. The product will be:

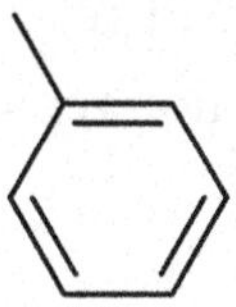

149

46.1. Iodination of Benzene

In order to place an iodo substituent on a benzene ring by EAS, we need a positively-polarized I. This is conveniently attained by using acetyl hypoiodite ($IOC(O)CH_3$). The mechanism is then the usual electrophilic addition/electrophilic elimination sequence:

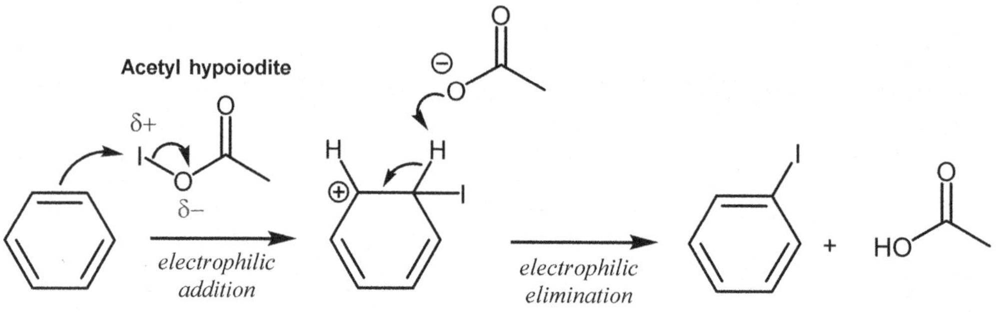

46.2. Chlorination and Bromination of Benzene

Unlike iodine, chlorine and bromine cannot be oxidized as easily to form free chlorenium or bromenium ions. When an electrophilic chlorine or bromine is needed for a reaction, a good alternative is to polarize the halogen by reaction with FeX_3 (here, X = Cl or Br):

Here, the dashed line indicates an attractive force that has not become a full bond and the X can be Cl or Br. The mechanism using this polarized halogen as the electrophile can be represented like this:

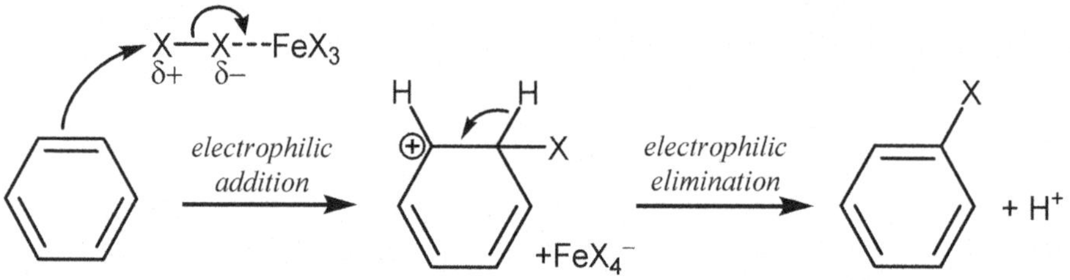

In practice, chlorination and bromination reactions may employ Fe powder instead of $FeCl_3$ or $FeBr_3$, respectively, because Cl_2 and Br_2 will react with Fe to make $FeCl_3$ or $FeBr_3$. So, there are a couple of ways that you might see this reaction represented:

150

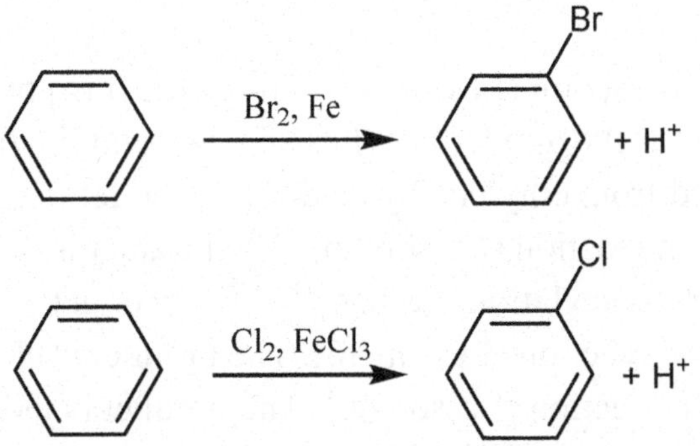

Br₂, Fe → bromobenzene + H⁺

Cl₂, FeCl₃ → chlorobenzene + H⁺

In the above examples, Cl_2 and Br_2 can be interchanged depending on whether one wants to add a Cl or Br to the benzene ring, respectively.

Example 46.1

Draw the major organic product for each of these reactions:

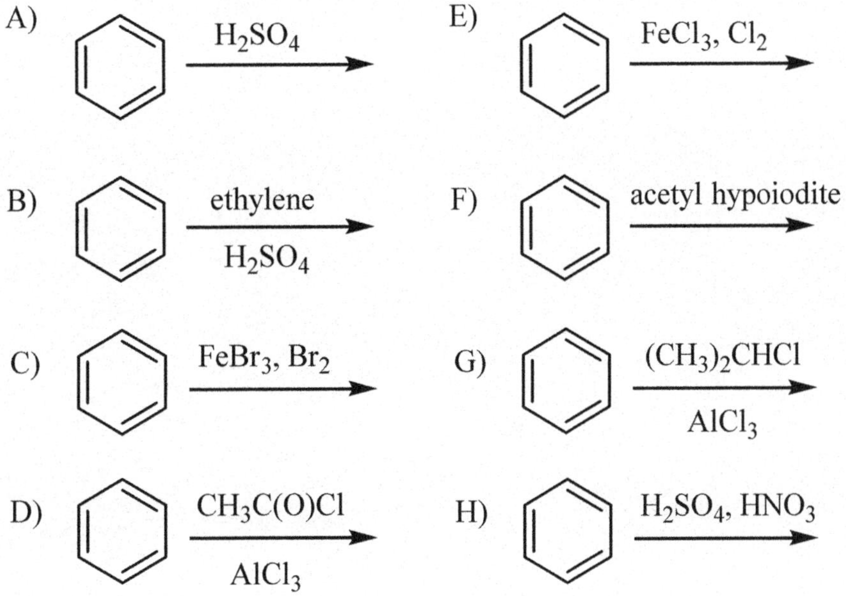

A) benzene $\xrightarrow{H_2SO_4}$

B) benzene $\xrightarrow[H_2SO_4]{\text{ethylene}}$

C) benzene $\xrightarrow{FeBr_3,\ Br_2}$

D) benzene $\xrightarrow[AlCl_3]{CH_3C(O)Cl}$

E) benzene $\xrightarrow{FeCl_3,\ Cl_2}$

F) benzene $\xrightarrow{\text{acetyl hypoiodite}}$

G) benzene $\xrightarrow[AlCl_3]{(CH_3)_2CHCl}$

H) benzene $\xrightarrow{H_2SO_4,\ HNO_3}$

<u>Solution 46.1</u>

A) This is a sulfonation reaction (Lesson 45). The product is phenylsulfonic acid ($PhSO_3H$).

B) This is a Friedel–Crafts alkylation (Lesson 44) and is the subset in which an ethyl carbocation is generated from ethylene by acid. The product is ethylbenzene.

C) This is a bromination reaction (Lesson 46). The product is bromobenzene.

D) This is a Friedel–Crafts acylation (Lesson 45). The product is acetophenone, ($PhC(=O)CH_3$).

E) This is a chlorination reaction (Lesson 46). The product is chlorobenzene.

F) This is an iodination reaction (Lesson 46). The product is iodobenzene.

G) This is a Friedel–Crafts alkylation (Lesson 45) and is the subset in which an isopropyl carbocation is generated from isopropyl chloride by $AlCl_3$. The product is isopropylbenzene.

H) This is a nitration reaction (Lesson 46). The product is nitrobenzene ($PhNO_2$).

47.1. More Stable Cations Form More Quickly in EAS

The key intermediate in EAS is the carbocation formed by electrophilic addition to the aromatic π-system. The formation of this carbocation is the rate-limiting step of the two-step sequence. In general, the more stable a carbocation is, the more quickly it will form, so anything that stabilizes the carbocation intermediate will accelerate the reaction, as illustrated by this reaction coordinate diagram:

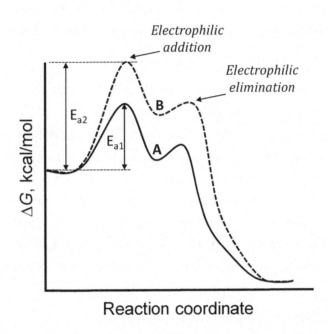

The energy of activation, E_{a1}, to form the more stable carbocation (intermediate **A**) is much lower than the energy of activation, E_{a2}, to form the less stable carbocation (intermediate **B**), so the EAS reaction proceeding through intermediate **A** will be faster.

47.2. Resonance Donor Substituents Activate the Ring to Faster Reaction

We know from Lesson 5 that resonance can stabilize carbocations. It thus follows that if a benzene ring has substituents that can donate electrons to stabilize the carbocation intermediate of an EAS reaction, the presence of these substituents will accelerate the reaction. Substituents with a lone pair-bearing atom directly adjacent to the benzene π-system will be the best resonance donors. For an atom to readily donate its lone pair into the π-system, however, that lone pair orbital must be comparable in size to the π-bond orbitals on C. To put it more succinctly, the lone pair-bearing atom should be from the same row as C (i.e., similar in size) to achieve optimum resonance donation. Electronegativity also plays a role in electron donating ability. Fluorine is so much more electronegative than carbon that it will not donate a lone pair via resonance. Taken together, this leaves us with O and N as elements that are in the same row as C and that commonly have lone pairs. The common **resonance electron-**

donating groups (EDGs) that will be **Activating for EAS** are –OR and –NR$_2$ (where R can be H or alkyl).

47.3. Alkyl Groups are Inductive Donors and Slightly Activate a Ring to EAS

We know from Lesson 25 that alkyl substituents can stabilize carbocations by hyperconjugation. It makes sense, then, that if a benzene ring has alkyl substituents which can stabilize the carbocation intermediate of an EAS reaction, the presence of these substituents will accelerate the reaction. Hyperconjugation is an example of an inductive effect, so **alkyl groups are classified as weakly inductive electron-donating groups (EDGs)**, which will be **Slightly Activating for EAS**.

47.4. Halogens are Weak Inductive Withdrawing/ Slightly Deactivating for EAS

Halogens substituents have lone pairs, but generally do not donate electrons to aromatic systems for two different reasons. Although fluorine is from the same row as C and will have an appropriate size match, it is significantly more electronegative than C, so it does not donate its lone pair electrons into a C-based π-system. Conversely, the heavier halogens are closer in electronegativity to C, but because they are so much larger, they are unable to overlap with the C-based π-system to donate electrons (i.e., a bad size match). Because each halogen is unable to participate in resonance donation (for different reasons), the inherent inductive electron-withdrawing effect of a halogens will dictate the rate of EAS for halogenated rings. Effects caused by nearby electronegative atoms are called inductive effect, so halide groups are classified as **weakly inductive withdrawing groups (EWGs)**, which will be **Weakly Deactivating for EAS**.

47.5. Strong Inductive Withdrawing Groups are Deactivating for EAS

When a substituent features a partial positive or formal positive charge on the atom immediately adjacent to the benzene ring, this charge will strongly repel any additional positive charge that would form on the aromatic system. As a result, this Coulombic repulsion means that a substituent with a partial positive or formal positive charge will significantly increase the energy of the carbocation intermediate formed during EAS. Substituents such as these are classified as **strongly inductive electron withdrawing groups (EWGs)**, which will be **Deactivating for EAS**. Common strong inductive withdrawing groups include –NO$_2$, –C(=O)R, –C(=O)OR, –C(=O)NR2, –C(=O)H, –CF$_3$, and –SO$_3$H.

47.6 Summary of Key Substituent Effects

For the purposes of this course, the key electron-donating groups to remember will be:

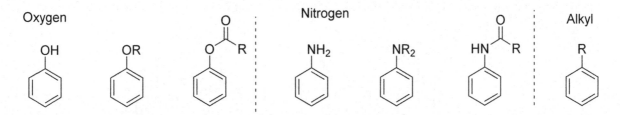

The key electron withdrawing groups will be:

$(X = Br, Cl, I, F)$

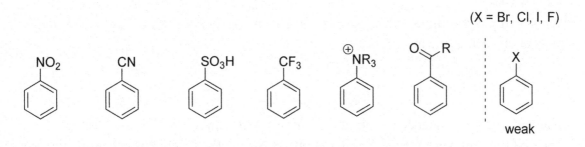

Example 47.1

Rank these substrates from fastest (1) to slowest (5) rate of electrophilic aromatic substitution reaction:

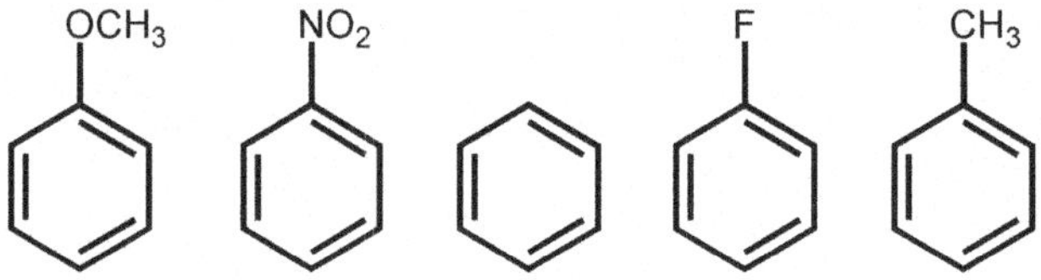

Solution 47.1

We start by determining what type of substituent (resonance donor, inductive donor, etc.) is present on each of the molecules.

From left to right:

The $-OCH_3$ is a resonance donor and is strongly activating.
The $-NO_2$ group has a formal 1+ charge on the N directly attached to the aromatic system. It is therefore strongly electron withdrawing and strongly deactivating
This is benzene without any substituents. This is the reference molecule or "baseline" to which each of the substrates is be compared.
The $-F$ is a halogen and is thus a weakly inductive withdrawing group that slightly deactivates the ring with respect to EAS.
The $-CH_3$ is a weak inductive donor and is slightly activating.

With these assignments in hand, we can now rank the molecules from fastest to slowest reaction rate in EAS:

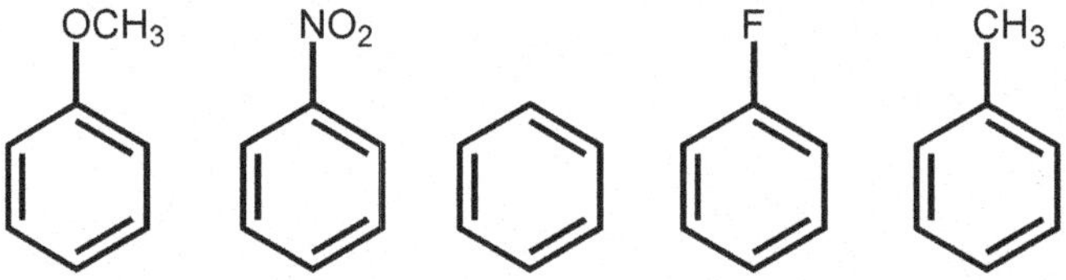

Lesson 48. Substituent Effects on the Regiochemistry of Electrophilic Aromatic Substitution

48.1. More Stable Intermediates Lead to Higher Product Yield

The placement of substituents on a benzene ring not only influences the rate of the EAS reaction (Lesson 47), but also on the regiochemistry of the products. What does this mean? When a monosubstituted benzene derivative undergoes EAS, for example, there are three possible sites where the electrophile can go relative to the substituent "Z", i.e., to form *o-*, *m-* and *p-* isomers:

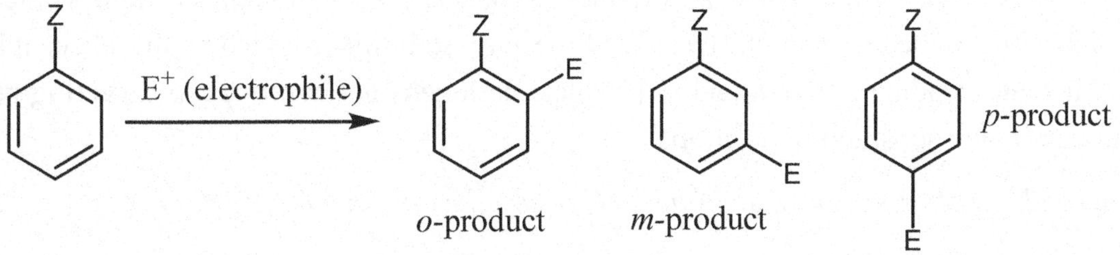

48.2. Activating Substituents are ortho-/para- Directors

Activating substituents (electron donors) can donate into an aromatic π-system. Electron donors push some negative charge onto the *o-* and *p-* positions, so E^+ (the electrophile) is attracted to those sites:

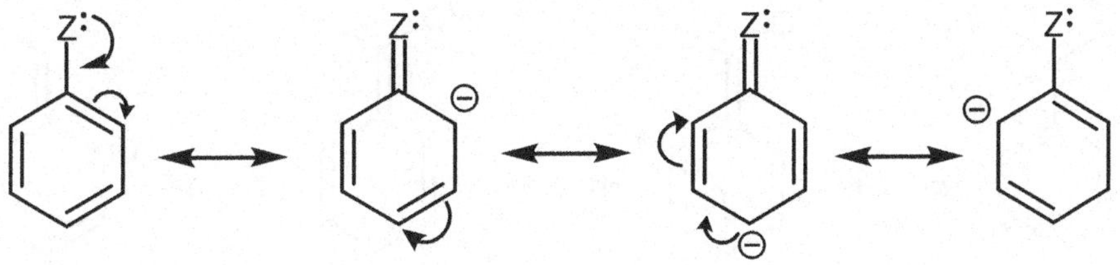

A similar argument can be made for inductive donors, which can stabilize carbocation intermediates leading to *o-* and *p-* substitution but nor those leading to *m-* substitution. The result is **that electron donors are *o-/p-*directors**. This means that the major products of EAS on a benzene that has an electron donor substituent will be those with the new substituent added *o-* or *p-* to the first substituent.

48.4. Weakly Deactivating Substituents (Halogens) are ortho-/para- Directors

Unlike resonance donors, halogen lone pairs (other than fluorine) are too poorly matched in size to effectively donate electrons to the benzene π system via resonance. Even fluorine, a good size match to C, does not effectively donate its lone pair by resonance. A lone pair adjacent to a cationic site, however, can stabilize a carbocation in a hyperconjugation-like fashion by through-space attraction between the lone pair and the empty *p*-orbital of a carbocation:

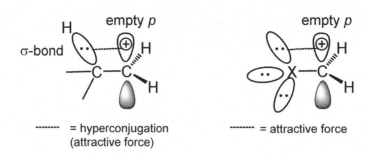

= hyperconjugation
(attractive force)

= attractive force

In the current context, the presence of this hyperconjugation-like stabilizing effect in the carbocation intermediates that lead to the *o*- and *p*- products makes these intermediates more stable than the carbocation leading to the *m*- product. Because of the extra stability provided by the adjacent lone pairs in the intermediates leading to the *o*- and *p*- products, **halogens are *ortho-*/*para*-directing groups** for electrophilic aromatic substitution reactions.

48.5. Moderately and Strongly Deactivating Groups are the only meta- Directors

The final class of substituents that we have to examine are moderately and strongly deactivating groups **(EWGs**, i.e., $-NO_2$, $-C(=O)R$, $-C(=O)H$, $-CF_3$, and $-SO_3H$). These types of substituents pull electrons out of the benzene ring and lead to formation of positive charges on the *o*- and *p*-positions. The result is that E^+ (the electrophile) is repelled from those sites and ends up favoring positions *m*- to the first substituent:

This leads to the fact that **electron-withdrawing groups are *m*-directing**. The *m*-substitution product will be the major product of EAS for a monosubstituted benzene ring having an EWG on it.

Example 48.1

Provide the major product(s) of chlorination of each of these starting compounds:

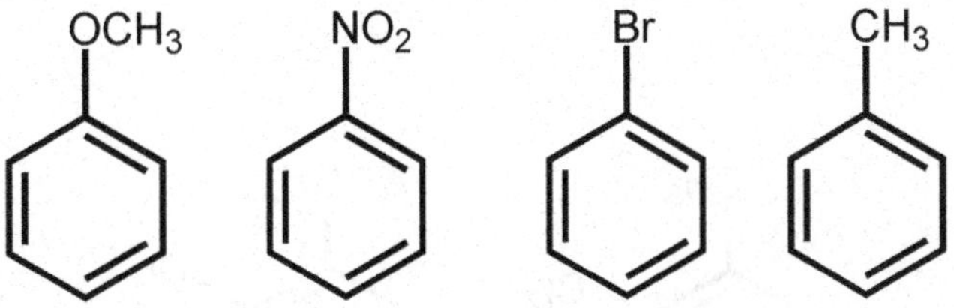

Solution 48.1

We start by determining what type of substituent (resonance donor, inductive donor, etc.) is present on each of the molecules.

From left to right:

The -OCH$_3$ is a resonance donor and is an *ortho-/para-* directing group.
The -NO$_2$ group is strongly electron withdrawing and *meta-* directing.
The -Br is a halogen, and these are weak inductive withdrawing groups that are *ortho-/para-* directing group.
The -CH$_3$ is a weak inductive donor and is an *ortho-/para-* directing group.

With these assignments in hand, we can now provide the major products of chlorination:

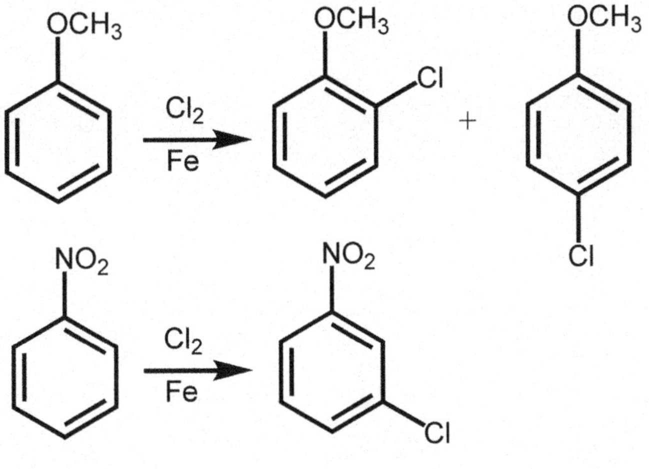

(cont'd)

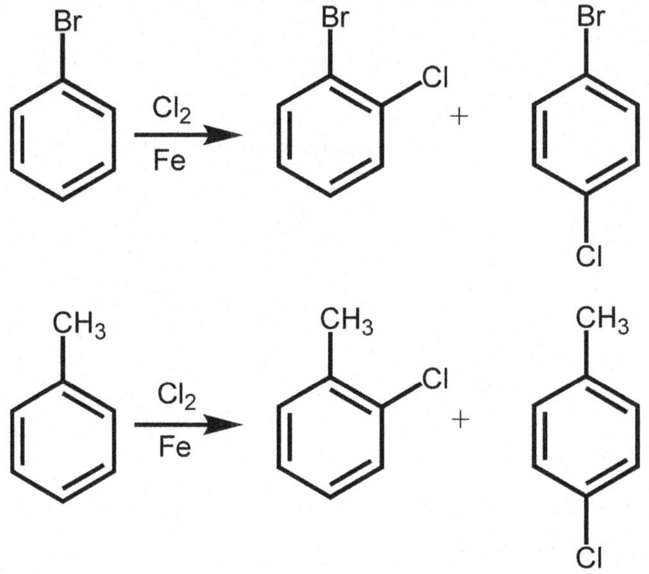
Br
Cl2
Fe
Br
Cl
+
Br
Cl
CH3
Cl2
Fe
CH3
Cl
+
CH3
Cl

Lesson 49. More on the Directing Effects of Substituents in EAS

49.1. EAS on Polysubstituted Benzene

In the previous Lesson, we saw that a substituent on a benzene ring influences the regiochemistry of the electrophilic addition step of a subsequent EAS reaction, directing the electrophile to add *o-/p-* or *m-* to the first substituent. What happens if *more than one* substituent is present on the benzene ring when we perform an EAS reaction? Consider an EAS on 2-ethylanisole. The $-OCH_3$ group would direct an electrophile to add to the sites indicated by the solid-line arrows in the diagram below:

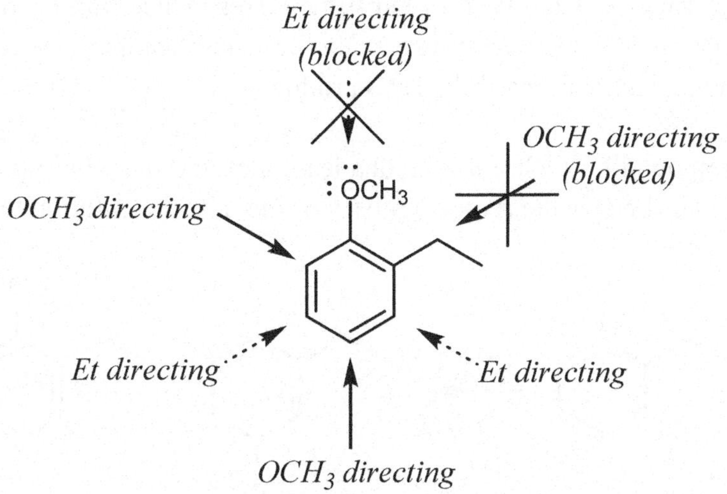

One of these sites is blocked because it already has a substituent on it (remember that an electrophile can only replace an H atom, not any other substituent, in EAS). On the other hand, the ethyl group would direct an electrophile to add to the sites *ortho-* or *para-* relative to the ethyl (indicated by the dashed-line arrows). As before, one of the sites is blocked (i.e., has a non-hydrogen substituent). So, how do we predict which site will actually be substituted in the major product when a disubstituted benzene undergoes EAS? The answer is surprisingly simple. The major product of a reaction is always derived from the most stable intermediate, therefore **the substituent that best stabilizes a carbocation intermediate will dictate the site of EAS regardless of how many substituents are on the ring**. From most-stabilizing to least-stabilizing categories of substituents, we have:

1) Resonance donors (i.e., $-OCH_3$, -OH, $-NR_2$, -OC(O)R)
2) Alkyl groups
3) Halogens (-F, -Cl, -Br, -I)
4) Electron-withdrawing groups (i.e., -C(O)R, $-CF_3$, $-NO_2$)

Example 49.1

Provide the major product(s) of Friedel-Crafts Alkylation (using CH_3Cl and $AlCl_3$) of each of these starting compounds:

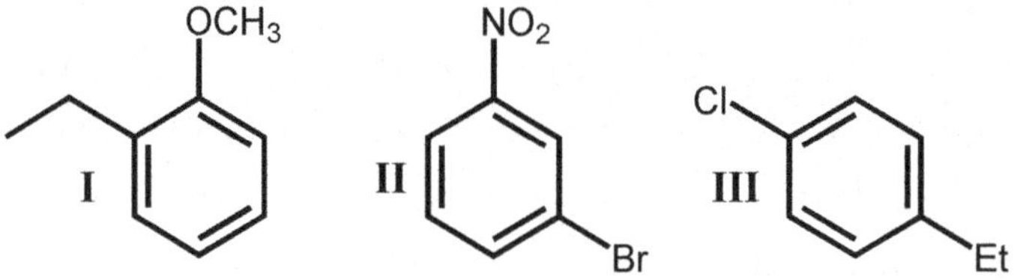

Solution 49.1

We start by determining what type of substituent (resonance donor, inductive donor, etc.) is present on each molecule and which type would best stabilize the carbocation intermediate (and thus dictates the regiochemistry of the EAS product).

For starting compound **I**, $-OCH_3$ is a resonance donor and the ethyl group is a weak inductive donor. The $-OCH_3$ dictates the regiochemistry of the next substitution to be *ortho-/para-* to the $-OCH_3$:

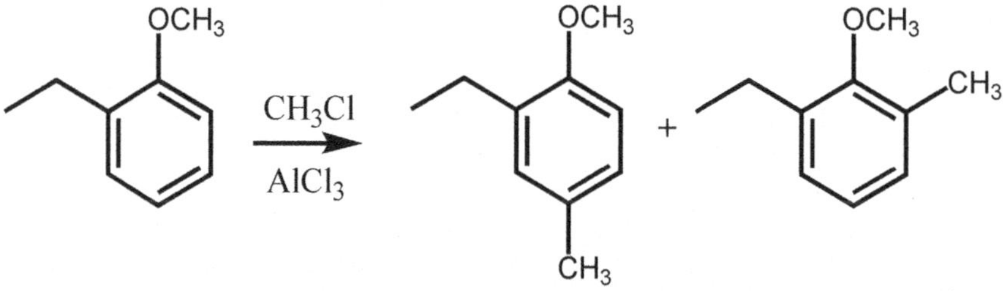

For starting compound **II**, the $-NO_2$ group is strongly electron withdrawing, whereas the $-Br$ is a weak inductive withdrawing group. The $-Br$ dictates the regiochemistry of the next substitution to be *ortho-/para-* to the $-Br$:

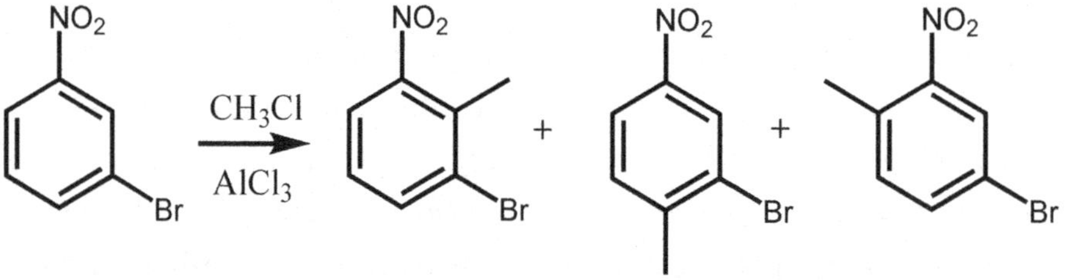

For starting compound **III**, the $-Cl$ is a weak inductive withdrawing group, whereas the Et group is a weak inductive donor. The $-Et$ dictates the regiochemistry of the next substitution to be *ortho-/para-* to the $-Et$. However, the *para* position is blocked by the Cl, and both *ortho* positions are chemically identical, so there is only one major product here:

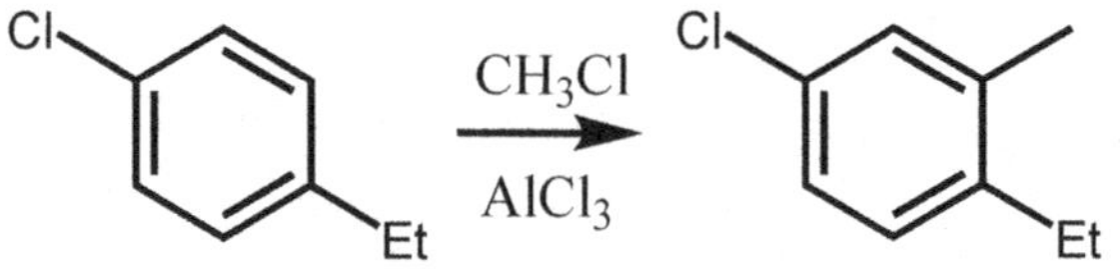

So far, we have only considered the electronic effects that contribute to carbocation stability, and we did not account for steric effects. Steric repulsion means that it is more difficult to place a new substituent *ortho-* to an existing substituent than it is to place a new substituent adjacent only to H atoms. The bigger a substituent is, the more steric strain it will exert on neighboring substituents. So, if at all possible (and it may not be possible), the major product will be the one that has the minimum amount of steric strain. However, steric effects do **not** generally supersede the electronic effects we learned for the directing ability of substituents. When predicting reaction products, you should always **first** identify the sites to which the electrophile will be directed and only then evaluate which of these "directed-to" sites has the least amount of steric strain.

Recall that, for EAS on a monosubstituted benzene, a number of substituents will direct the electrophile to yield a mixture of *o-* and *p-* products. This may seem to contradict the idea of minimizing steric strain, which should favor making exclusively *para*-products. After all, the *o-* product places the second substituent immediately adjacent to the first, whereas the *p-* product places the second group as far from the first as possible. The observation that a mixture of *o-* and *p*-products are formed can be explained by the fact that there are *two* possible *ortho-* substitution sites but only *one* possible *para*-substitution site. So, statistics would favor formation of the *ortho-* product, but sterics we would favor formation of the *para-* product. These two factors balance each other out (roughly speaking), which affords a mixture of the *ortho-* and *para-* products. An exception would be if one of the two substituents in the product is bulky (has three non-H branches off of the point of attachment to the benzene ring, for example). In such cases we will get predominantly the *p-* product.

Finally, it is generally quite a bit more difficult to place a substituent between two substituents than next to one substituent with an H on the other side. We should consider such sites only when we have no other choice based on the directing ability of the existing substituents.

Example 49.2

Provide the major product(s) of nitration of each of these starting compounds:

Solution 49.2

For molecule **I**, –OCH₃ is a resonance donor and the *t*-butyl group is a weak inductive donor. The –OCH₃ dictates the regiochemistry of the next substitution to be *ortho-/para-* to the –OCH₃. Two of these sites are directly adjacent to the bulky *t*-butyl group (has three non-H branches), so steric effects will strongly disfavor the electrophile reacting at these sites. The major product, then, is the only one in which substitution is *both ortho-* to the -OCH₃ and not directly adjacent to the bulky *t*-butyl group:

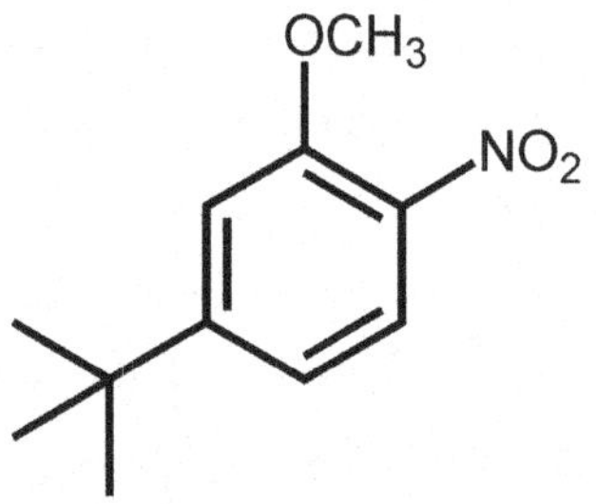

For molecule **II**, the –NO₂ group is strongly electron withdrawing, the –Cl is a weak inductive withdrawing group, and the isopropyl is a weak inductive donating group. The isopropyl dictates the regiochemistry of electrophilic addition to be *ortho-/para-* to itself. One *ortho-* position is already occupied by a Cl. The other *ortho-* site is between the isopropyl and the Cl, so steric effects disfavor electrophilic addition at that. As a result, substitution *para-* to the isopropyl is the major product:

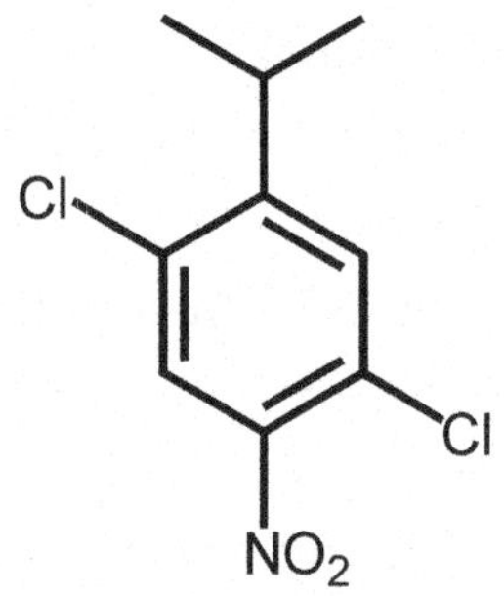

Lesson 50. Preparation of Carboxylic Acids

Lesson 50.1. Oxidation of Alcohols Using Chromium Reagents

In Lesson 28, we saw that H^+/CrO_4^{2-}, $H^+/Cr_2O_7^{2-}$, and CrO_3/H_2SO_4 (reagents for what is called the **Jones Oxidation**) are stronger oxidants than are PCC or PDC. Since these reagents are powerful, they will transform primary alcohols into carboxylic acids by further oxidizing aldehydes. These oxidizing reagents can be used to produce ketones with secondary alcohols, but aldehydes cannot be produced using these reagents.

Example 50.1

Give the products of the following reaction:

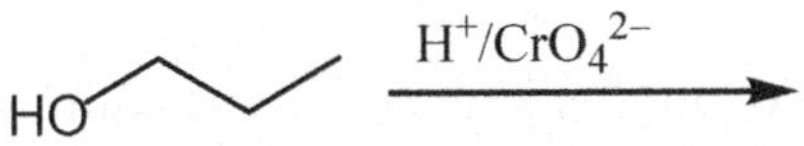

Solution 50.1

These oxidizing agents are strong enough to mediate oxidation of all carbinol H atoms. This leads to conversion of primary alcohols to carboxylic acids: This leads to conversion of secondary alcohols to ketones and of primary alcohols to carboxylic acids:

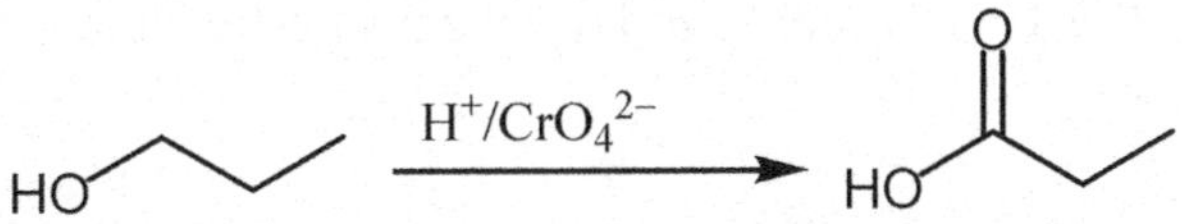

Lesson 50.2. Ozonolysis of Alkenes

As we saw in Lesson 34, alkenes react with ozone in an oxidation reaction where the C=C bond breaks and is replaced with two C=O bonds. This reaction necessitates a second step which can be an oxidation step (usually H_2O_2), where ketones and carboxylic acids can be produced (depending if we have a hydrogen on the alkene or not).

Example 50.2

Give the products of the ozonolysis of 2-methyl-2-butene followed by both oxidation step

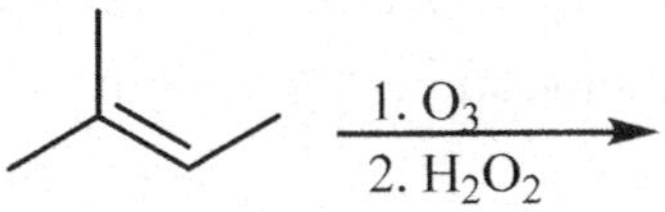

Solution 50.2

In these reactions, the C=C breaks and two carbonyl molecules are formed. Since we have ozonolysis followed by an oxidation step, the products are going to be a ketone and carboxylic acid, since carbon #3 has a hydrogen on it and the molecule will further oxidize.

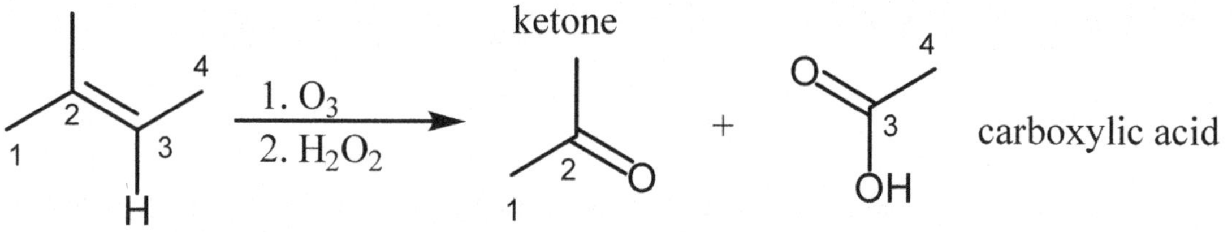

ketone

carboxylic acid

Lesson 50.3. Ozonolysis of Alkynes

When alkynes react with ozone in an oxidation reaction where the C≡C bond breaks and is replaced with two C=O bonds. Regardless of the second step, whether it is a reduction step or oxidation step, the product is a carboxylic acid.

Example 50.3

Give the products of the ozonolysis of 2-pentyne followed by water.

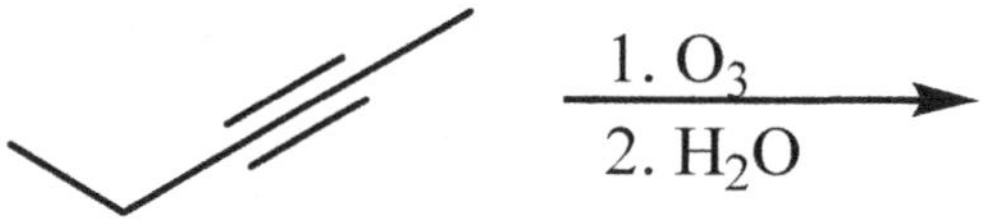

Solution 50.3

In this reaction, the C≡C breaks and two carboxylic acids are formed, an acetic acid and propanoic acid.

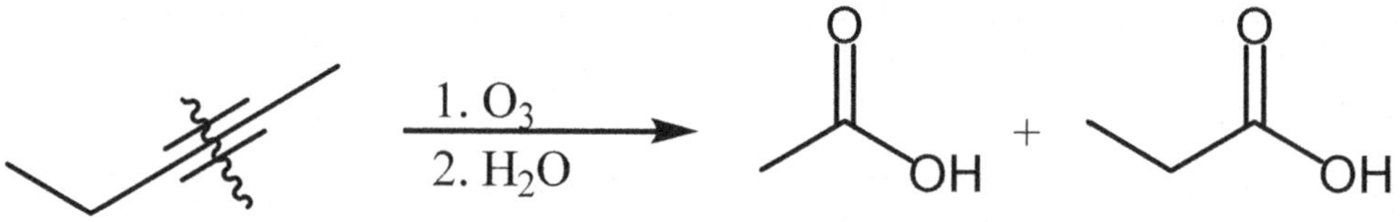

Lesson 50.4. Hydrolysis of Nitriles

We have seen in Lesson 23 that primary and secondary alkyl halides can react with CN⁻ (cyanide ions) in S_N2 reaction mechanism to give nitriles (R–CN). Treating nitriles with water in presence of an acid catalyst and heat will give a carboxylic acid. This reaction is called "hydrolysis reaction". The mechanism of this reaction is not included in this Lesson.

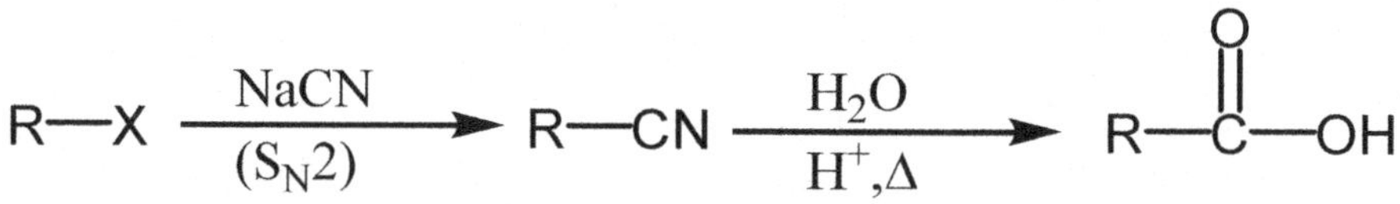

The significance of this reaction is if we start with an alkyl halide with certain number of carbons, upon substituting it with CN⁻ followed by hydrolysis, a carbon will be added to the main chain as we can see from the following example:

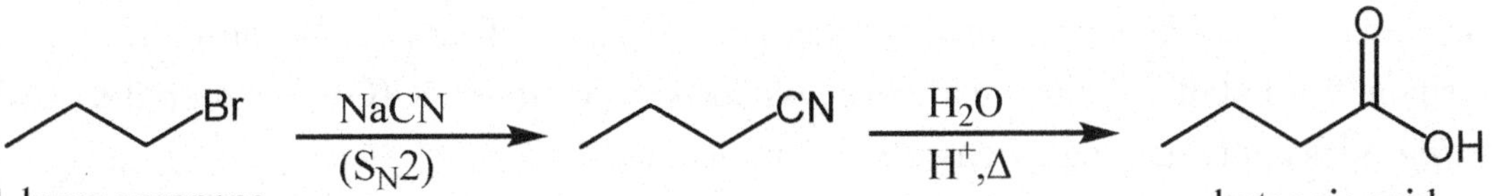

1-bromopropane
3 carbon chain

butanoic acid
4 carbon chain

Starting with three-carbon chain (1-bromopropane) we can make a four-carbon chain (butanoic acid). So, we always keep in mind when we are using this synthesis route for carboxylic acids that we are adding one more carbon to the main chain.

Lesson 51.1. Naming Carboxylic Acids

Nomenclature rules for carboxylic acids again mirror the rules for naming alkanes, but for carboxylic acids we use *-oic acid* in place of the *-e* at the end of the name. Because the carboxylic acid is necessarily at the end of a chain, we always give the carbon of the –C(O)OH group the number 1. All other *naming* rules apply as usual. Consider this molecule.

(R)-3-hydroxy-5-methylhexanoic acid

The carbon in the –C(=O)OH group must be assigned "1", and the –OH group on carbon 3 will be listed as a "hydroxy" substituent (a carboxylic acid takes priority over an alcohol). The remaining substituents and stereochemistry are determined using the standard conventions.

Lesson 51.2. Naming Esters

Naming an ester is slightly more involved than naming a carboxylic acid. When naming esters, we first list the name of the alkyl group "R–" that is attached to the oxygen. We then name the main chain (R"–C(O) portion), assigning the C of the –C(O)O unit as "1" and substituting the "-e" in the suffix by "-oate". The figure below shows the general formula for naming the ester.

main chain ending with "*oate*" ⟵

Consider a specific example.

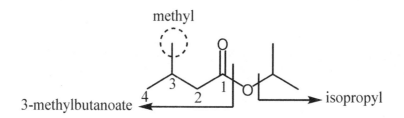

3-methylbutanoate ⟵

In this molecule, the alkyl group on the oxygen is an isopropyl group, the main chain (starting with the carbonyl carbon) has four carbons, and carbon "3" has a methyl substituent. Therefore, the main chain is 3-methylbutanoate. The name of this ester will be given as **isopropyl 3-methylbutanoate** (there must be a space between the oxygen substituent and the ester chain names).

Lesson 52. Nucleophilic Addition of Organometallics and Metal Hydrides

Lesson 52.1. Introduction to Reaction Mechanism

When treating carboxylic acids and their derivatives with some nucleophiles, a nucleophilic substitution step may be followed by a nucleophilic addition then protonation.

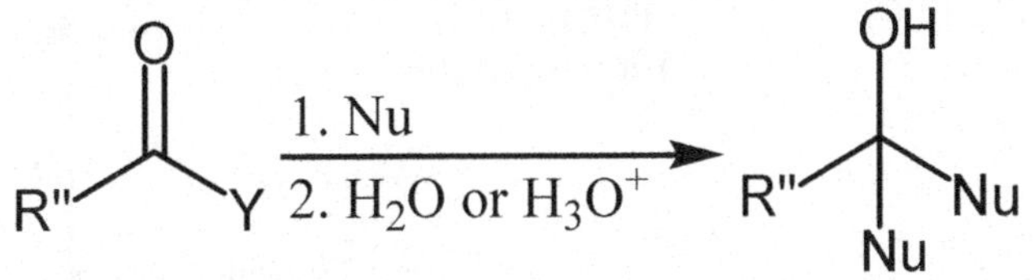

The general mechanism for these reactions is.

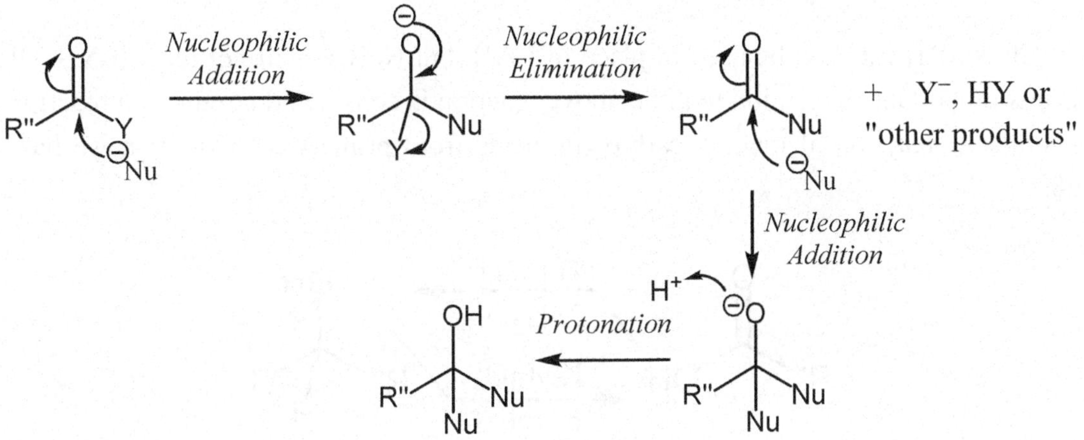

The following Lessons will cover the reactants/reagents that facilitate these reactions

Lesson 52.2. Reactions with $LiAlH_4$

When carboxylic acids, acid chlorides, or esters react with strong reducing agents (hydride donors like $LiAlH_4$, abbreviated LAH), the nucleophilic substitution reaction forms the aldehyde. However, a strong hydride donor (e.g. $LiAlH_4$ and few other reagents) will react further by a nucleophilic addition/protonation. The following figure illustrates the general mechanism of this reaction.

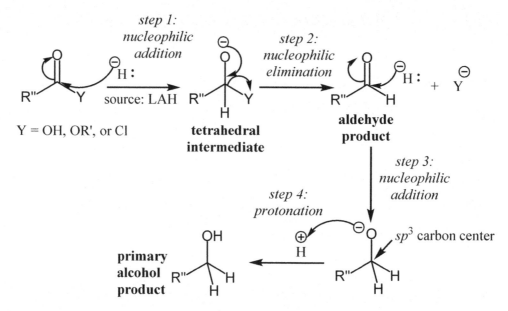

If Y = Cl, NaBH₄ can also be used in place of LAH, but NaBH₄ will not react if Y = OH, OR or NR₂. It is also important to mention that the above reaction is considered a reduction reaction and it is the exact opposite reaction of making carboxylic acids from primary alcohols that we have seen in Lesson 28.

Example 52.1

Give the products of the following reaction.

Solution 52.1

The reaction of LAH with carboxylic acids followed by hydration will give a primary alcohol. Note that the stereochemistry of any carbon on the main chain will not change.

Lesson 52.3. Reactions with RMgX (X=Cl, Br, I) or RLi

Treating esters and acid chlorides with Grignard reagents (RMgX, X is Cl, Br, or I) or organolithium reagents (RLi) will give a tertiary alcohol with two new σ C–C bonds. The reaction mechanism is same as Lesson 52.2, but with R⁻ in place of hydride. The general mechanism is as follows.

The reactivity of RMgX and RLi diverge in the case of carboxylic acids. Keep in mind that Grignard and organolithium reagents are still strong bases, so these reagents will first deprotonate a carboxylic acid to generate a carboxylate (RCO_2^-) before any nucleophilic addition is able to occur. This negative charge makes the carbonyl group significantly less susceptible to nucleophilic addition in a carboxylate relative to a carboxylic acid. Consequently, the R⁻ in RMgX is not sufficiently nucleophilic to add to the carbonyl in a carboxylate, and the reaction stops at this point.

Example 52.2

Give the products of the following reaction.

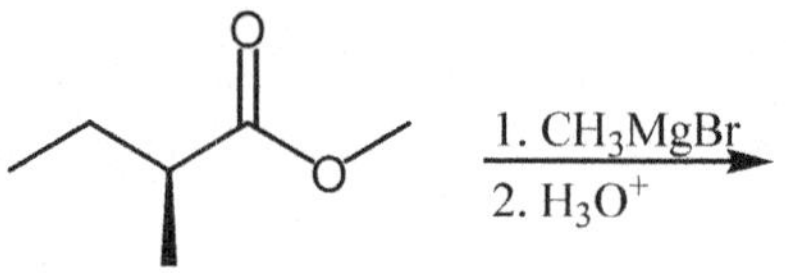

$$\xrightarrow[\text{2. } H_3O^+]{\text{1. } CH_3MgBr}$$

Solution 52.2

The addition of CH_3–MgBr (we can think about it as CH_3^-) to ester will give a tertiary alcohol with the two methyl groups attached to the carbon, and ethanol will be a byproduct of the reaction. Here is a simple mechanism to explain the products of the reaction.

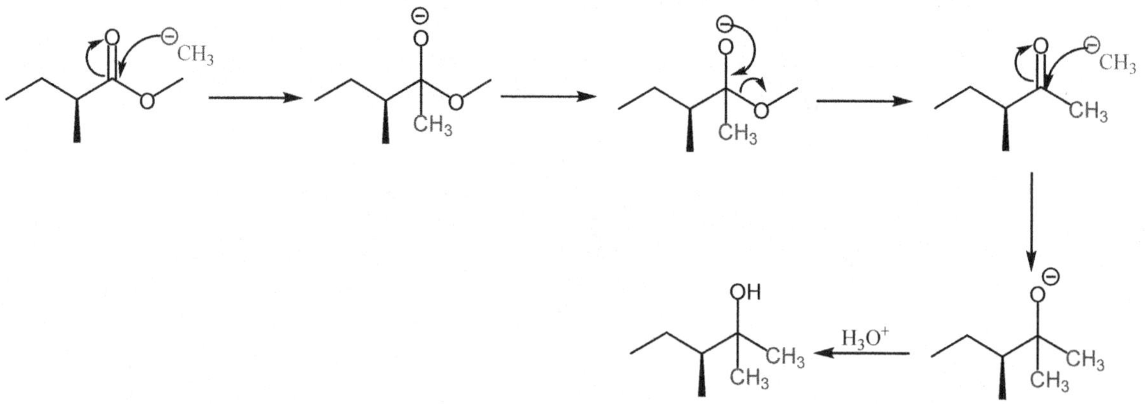

Lesson 53. Introduction to Nucleophilic Acyl Substitution

Lesson 53.1. Introduction to Nucleophilic Acyl Substitution

Carboxylic acids and their derivatives (anhydrides, acid chlorides, amides and esters) undergo **nucleophilic acyl substitution**.

Nucleophilic acyl substitution is a two-step process:

1) Nucleophilic addition to the carbonyl carbon and

2) Nucleophilic elimination of the best leaving group from what was the carbonyl carbon.

tetrahedral intermediate

Lesson 53.2. Carboxylic Acid Derivatives Have Different Reactivity to Nucleophilic Addition.

The different carbonyl-based functional groups have different reactivity towards nucleophilic addition, and it is important to understand the origins of these differences in reactivity. For example, treating acetyl chloride with NH_3 (the nucleophile) will afford the amide, but treating the amide with HCl (Cl^- is the nucleophile) does not produce acetyl chloride, even though the anionic Cl^- is a better nucleophile than the neutral NH_3.

If we recall from Lesson 7 that the stronger the acid the more stable the conjugate base. So, to compare the stability of "Y^-" in the carboxylic acid derivatives we compare the acidity of "HY", the conjugate acid (C.A.). The table below gives the pK_a values for the C.A. of the different "Y^-" and thus the stability of the leaving groups.

Y^-	Conjugate acid	pK_a of C.A.	Leaving group ability	Reactivity
Cl^-	HCl	-7	Best leaving group	Most reactive
$RC(O)O^-$	$RC(O)OH$	5		
RO^-	ROH	15		
NR_2^-	NR_2H	35	Worst leaving group	Least reactive

We can see from the table that the best leaving group is the chloride ion, which makes acid chlorides the most reactive carboxylic acid derivatives, after which the anhydrides and then the esters and the least reactive carboxylic acid derivatives are the amides.

So, in conclusion, when comparing the reactivity of carboxylic acid derivatives, we can think of the following trend:

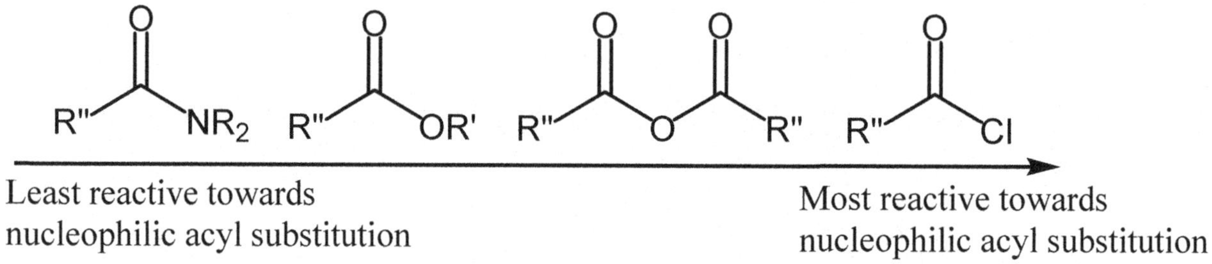
R''
NR2
R''
OR'
R''
O
R''
R''
Cl
Least reactive towards nucleophilic acyl substitution
Most reactive towards nucleophilic acyl substitution

Lesson 54. Nucleophilic Acyl Substitution of Carboxylic Acids to form Acid Chlorides

Carboxylic acids react with phosphorus trichloride (PCl_3) or thionyl chloride ($SOCl_2$) to give acid chlorides. In these cases, the –OH is substituted for a –Cl:

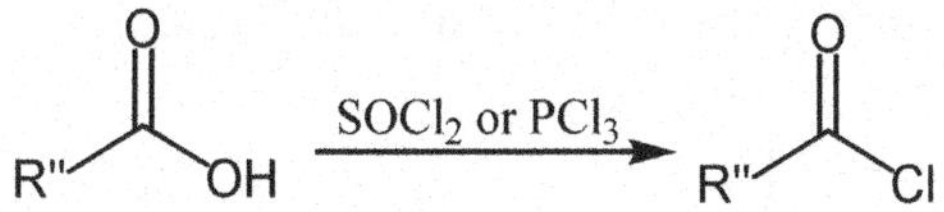

This nucleophilic acyl substitution is initiated by the reaction of the carbonyl oxygen with the sulfur of $SOCl_2$ or the phosphorus of the PCl_3, generating the chloride ion that will act as a nucleophile:

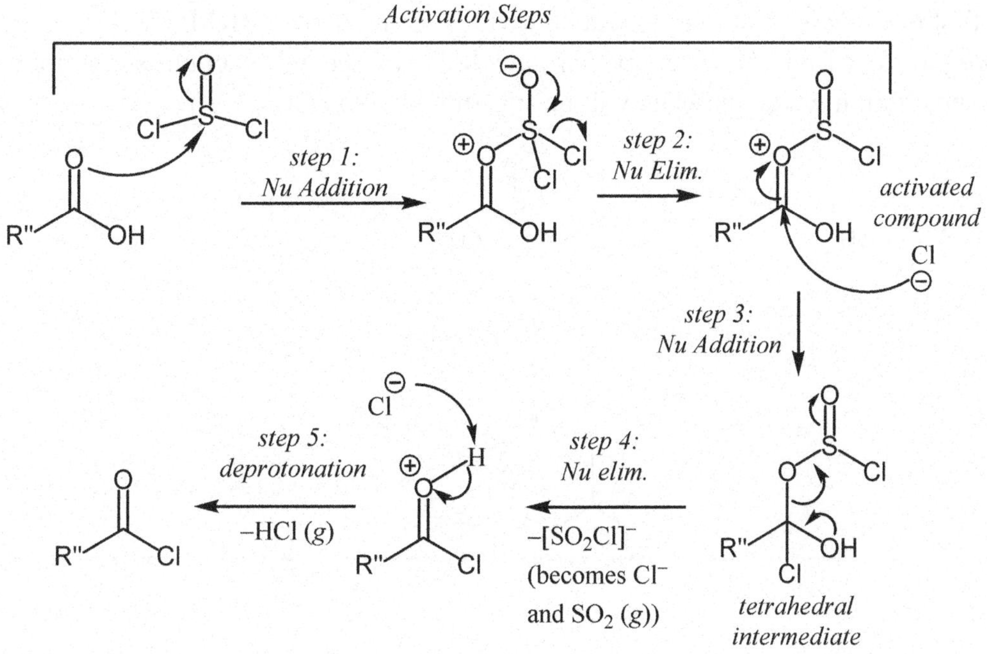

Here, steps 1 and 2 are needed to activate the complex (the OH must be turned into a good leaving group), and to produce the Cl^- needed for subsequent nucleophilic acyl substitution. These two steps represent a nucleophilic acyl substitution to an S=O group, in which the carbonyl O acts as a nucleophile to add to the sulfur in the S=O bond. Steps 3-4 represent nucleophilic substitution, then nucleophilic elimination. Step 5 is needed to deprotonate the product of nucleophilic acyl substitution to yield a neutral product. The production of sulfur dioxide and HCl as side products makes the purification of this reaction rather easy, because both of these side products are gases that bubble out of the reaction vessel and can be collected.

The reactions of carboxylic acids with PCl_3 proceed through a very similar mechanism to that observed for reaction with $SOCl_2$.

Example 54.1

Give the final product of the following two-step sequence:

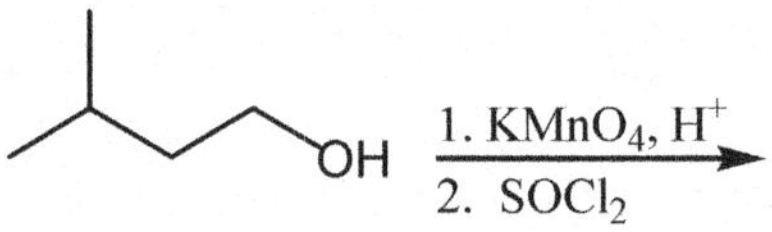

$$\xrightarrow[\text{2. SOCl}_2]{\text{1. KMnO}_4, \text{H}^+}$$

Solution 54.1

The first reaction is an oxidation reaction with a strong oxidant ($KMnO_4$), which will produce a carboxylic acid from the given primary alcohol. The second reaction is a simple nucleophilic acyl substitution where the –OH will be substituted with –Cl

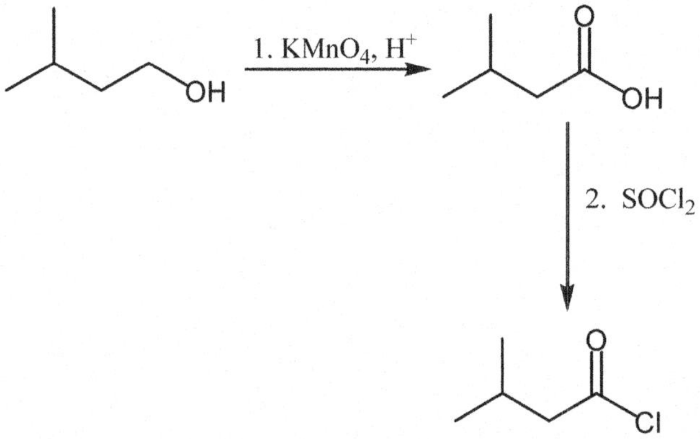

Lesson 55. Nucleophilic Acyl Substitution of Carboxylic Acids to form Esters

Carboxylic acids react with alcohols in presence of an acid catalyst (such as H_2SO_4) and heat to produce esters (and water as a byproduct), this reaction is known as **Fischer Esterification**. The general reaction is the following:

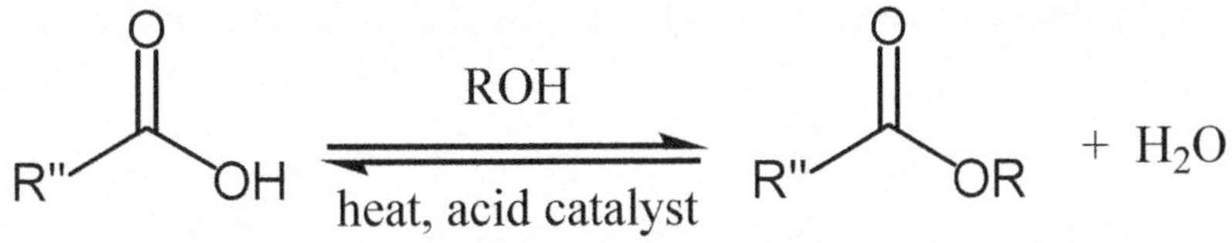

The following figure shows the arrow pushing and classification for each of the 6 steps in Fischer esterification mechanism.

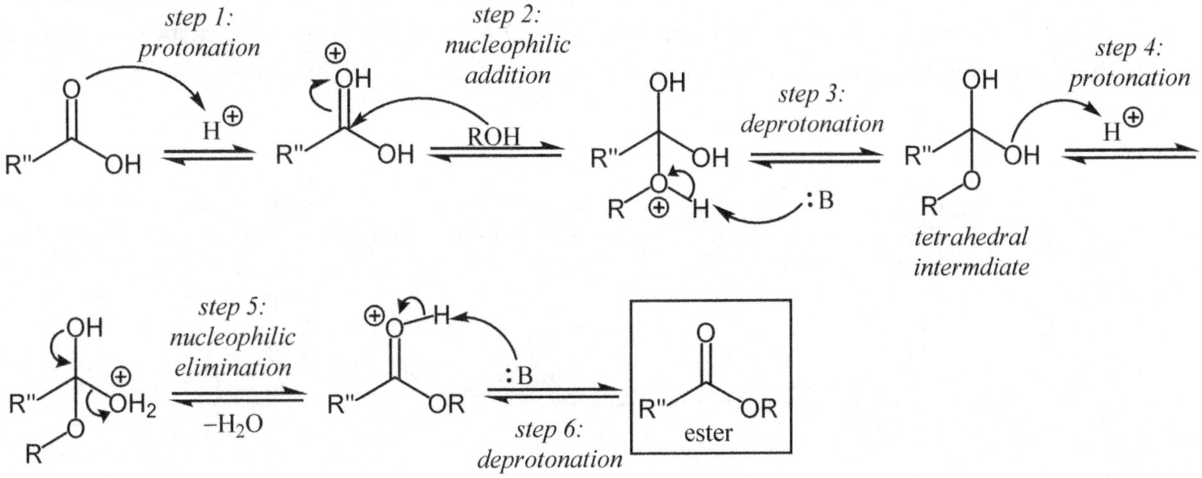

Step 1: the acid catalyst (simplified as H^+) protonates the carbonyl oxygen and generates the intermediate activated towards nucleophilic attack

Step 2: the alcohol acts as the nucleophile for nucleophilic addition to the activated carbonyl species, breaking the C–O π-bond and forming a new C–O σ-bond

Step 3: a base in solution (another alcohol molecule, the conjugate base of the acid catalyst) removes a proton to form the neutral tetrahedral intermediate

Step 4: an OH group in the tetrahedral intermediate is protonated by the acid catalyst

Step 5: H_2O eliminates, which produces a C–O π-bond and reestablishes the sp^2 carbon center

Step 6: a base in solution deprotonates the oxygen to yield the neutral ester product

Note that **each step of the reaction is reversible**. The ester product can react with the water and hydrolyze to give back the carboxylic acid, a useful reaction called **ester hydrolysis**. To ensure the production of the ester in high yields we will apply our knowledge of equilibrium processes in terms of LeChatelier's principle. Specifically, we can use a large excess of the alcohol reagent so that its

177

consumption pushes the equilibrium towards ester formation. Alternatively, we can distill the water away as it forms, again to push the reaction towards ester formation.

Example 55.1

Isoamyl acetate (structure below) is an ester found in banana oil and is responsible for the familiar banana aroma. Give the carboxylic acid and alcohol used in the formation of this ester.

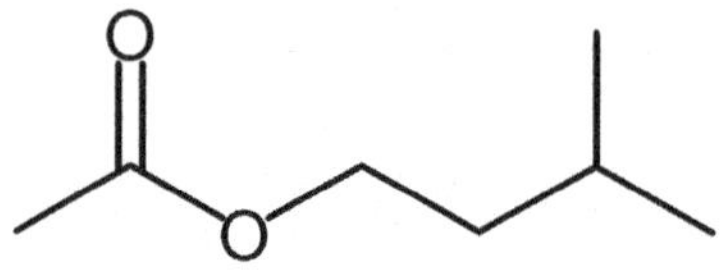

Solution 55.1

In order to find the reactants of this ester we put the general reaction and try to find both R" and R of the carboxylic acid and alcohol respectively. As the figure shows the carboxylic acid used is the acetic acid and isoamyl alcohol (3-methyl-1-butanol) is the alcohol used.

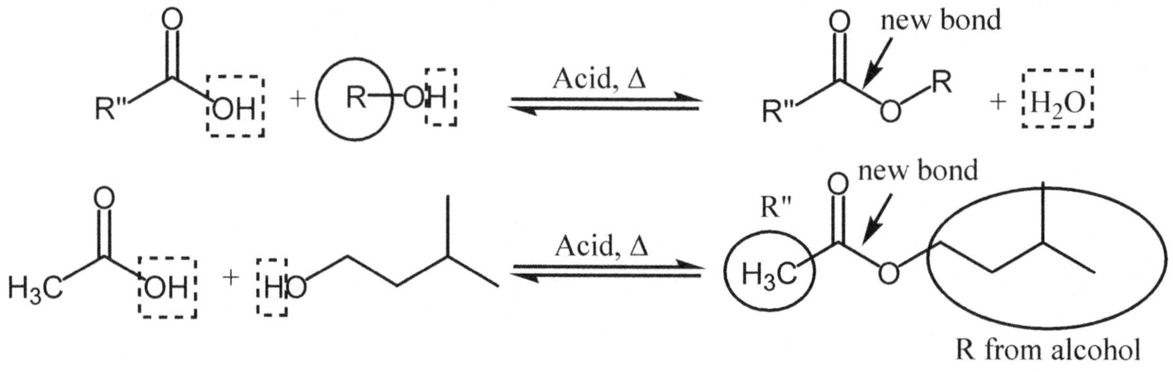

Example 55.2

Provide the missing reactants and products:

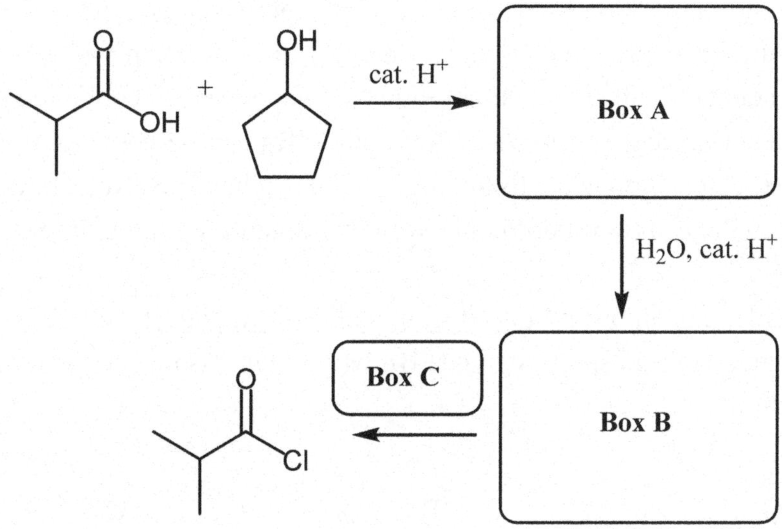

Solution 55.2

The product in Box A will be an ester, formed via Fischer Esterification. Reaction of the Ester with water/acid catalyst takes us back to the carboxylic acid in Box B. The missing reagents in Box C must facilitate conversion of the carboxylic acid into an acid chloride. Either $SOCl_2$ or PCl_3 will accomplish this. The completed scheme would be:

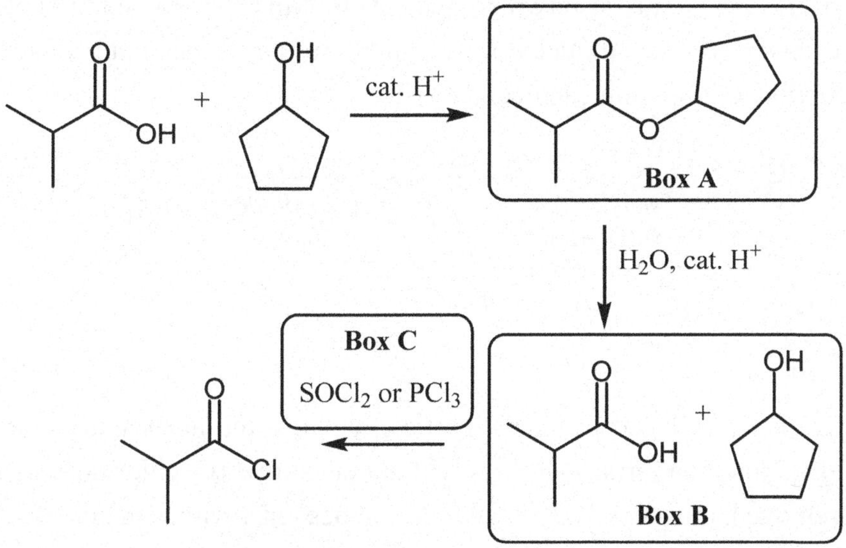

Lesson 56. Nucleophilic Acyl Substitution of Acid Chlorides and Anhydrides

Lesson 56.1. Nucleophilic Acyl Substitution of Acid Chlorides

As we saw in Lesson 55, acid chlorides are the most reactive carboxylic acid derivatives. In order to make any carboxylic acid derivative, a good strategy is thus to begin with the carboxylic acid and then convert it to an acid chloride. The acid chloride can then be easily converted to any of the desired derivatives via nucleophilic acyl substitution. Acid chlorides can be converted to anhydrides, esters, amides, and carboxylic acids through the following general mechanisms. Note, however, that if a neutral nucleophile is used it will need to be deprotonated to give a neutral product in the end.

Anionic nucleophile (e.g. anions produced from NaOC(O)R, NaOH, NaOR, etc.)

This is a nucleophilic acyl substitution (Type B), where step 1 is nucleophilic addition and step 2 is nucleophilic elimination:

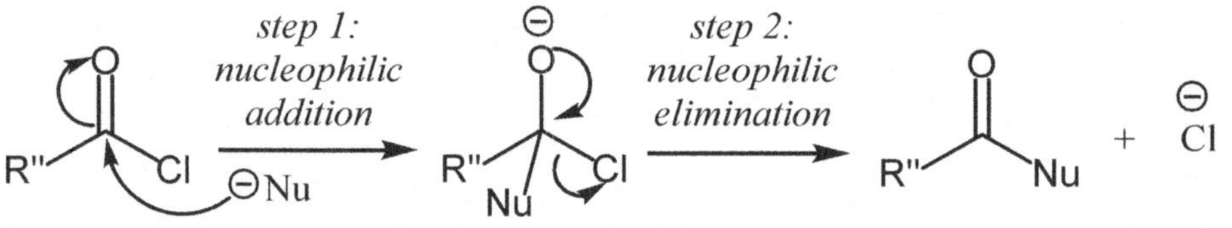

Neutral nucleophile (e.g. ROH, NR_2H, NRH_2, NH_3 or H_2O)

This mechanism generally requires the presence of base catalyst. If the NuH is itself a base (NuH = NR_2H, NRH_2 or NH_3), however, then we just use two equivalents of NuH (one to be the nucleophile, one to be the base). This mechanism consists of nucleophilic addition first as usual, but a deprotonation step is required before nucleophilic elimination can occur:

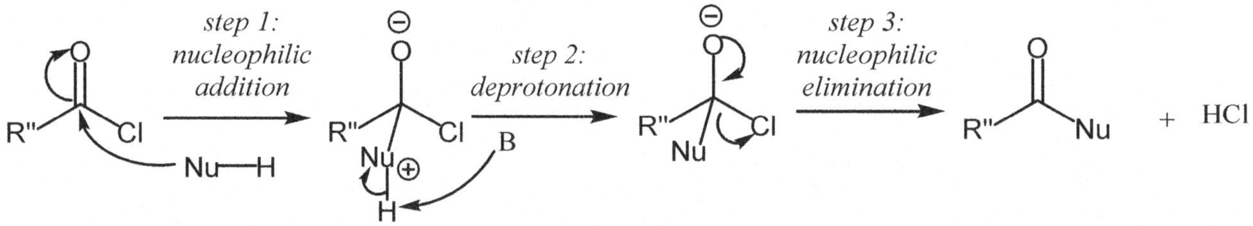

The carboxylic acid derivatives produced by nucleophilic acyl substitution of acid chlorides are summarized in the chart below. Note that tertiary amines will not substitute the chloride because they lack the proton that must be removed, thus only NH_3, 1°, and 2° amines react with acid chlorides to give amides:

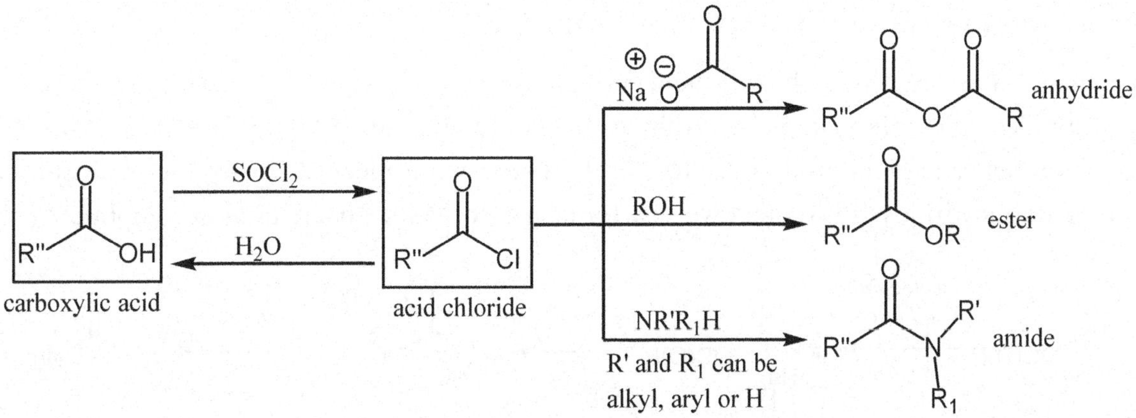

SOCl$_2$

H$_2$O

R"—COOH carboxylic acid

R"—COCl acid chloride

Na$^{\oplus}$ $^{\ominus}$O—CO—R

anhydride

ROH

ester

NR'R$_1$H

R' and R$_1$ can be alkyl, aryl or H

amide

Example 56.1

Give the final product of the following sequence of three reactions:

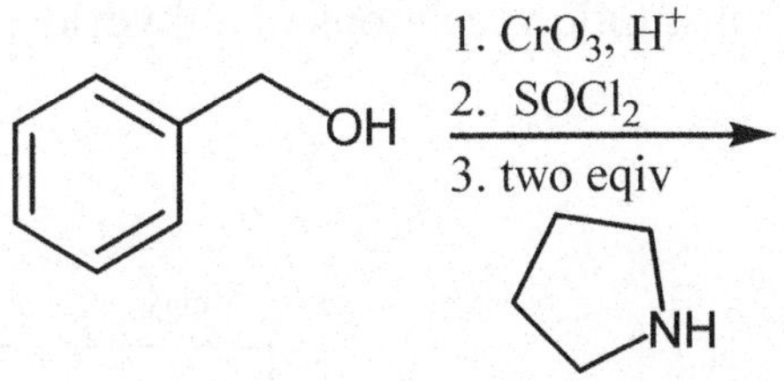

1. CrO$_3$, H$^+$
2. SOCl$_2$
3. two eqiv

pyrrolidine (NH)

Solution 56.1

The first step is an oxidation reaction with a strong oxidant (CrO$_3$, H$^+$) which will produce a carboxylic acid from the given primary alcohol. The second step is a simple nucleophilic acyl substitution where the C—OH is substituted with C—Cl. The third step is an **amidation reaction** where a C—Cl is substituted with C—NR$_2$:

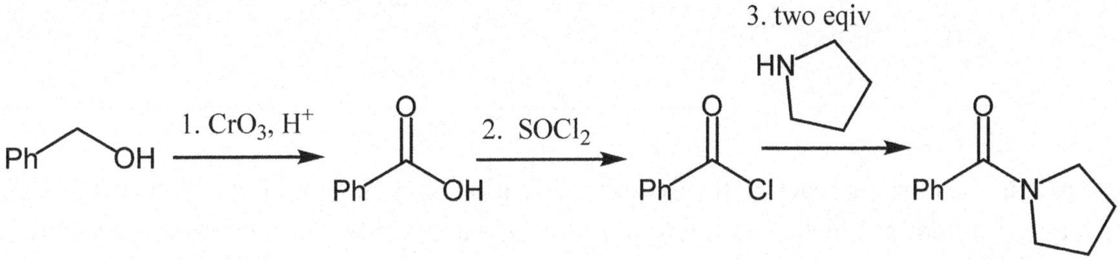

3. two eqiv

1. CrO$_3$, H$^+$

Ph—CH$_2$—OH

Ph—COOH

2. SOCl$_2$

Ph—COCl

HN (pyrrolidine)

Ph—CO—N (pyrrolidine)

Lesson 56.2. Nucleophilic Acyl Substitution of Anhydrides

The second-most reactive carboxylic acid derivatives are acid anhydrides, and they can be converted to esters, amides and carboxylic acids. Anhydrides, however, cannot be converted to acid chlorides, since Cl⁻ is a better leaving group than RC(O)O⁻. The general mechanism for the nucleophilic acyl substitution of anhydrides follows that we saw for acid chlorides with a neutral nucleophile:

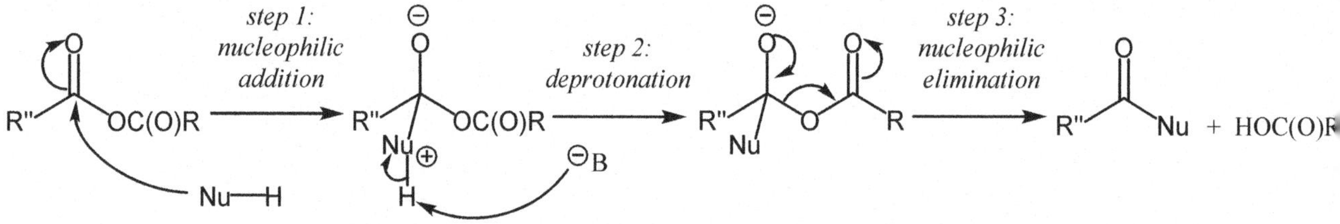

In these cases, Nu–H can be H_2O, ROH, a 1° amine, or a 2° amine. Regardless of the particular nucleophile, the first step is a nucleophilic addition reaction, the second step is deprotonation of Nu and the third step is nucleophilic elimination of RC(O)O⁻ (a better leaving group than Nu⁻). The following chart shows the nucleophilic acyl substitution reactions of anhydrides that give rise to carboxylic acids, esters, and amides.

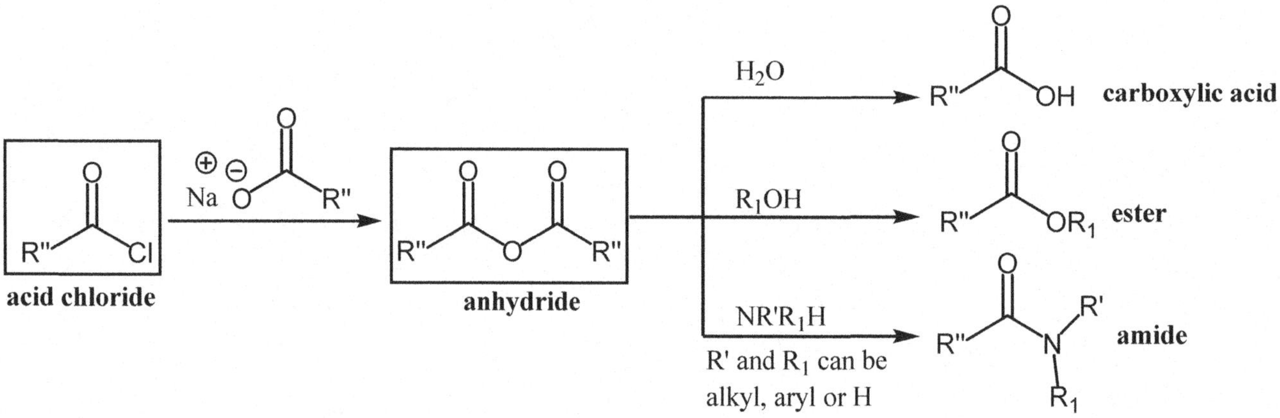

Example 56.2

Provide the missing reactant for the transformation shown. Would the proposed product be the major or minor product upon reaction of this acid anhydride with the missing reactant?

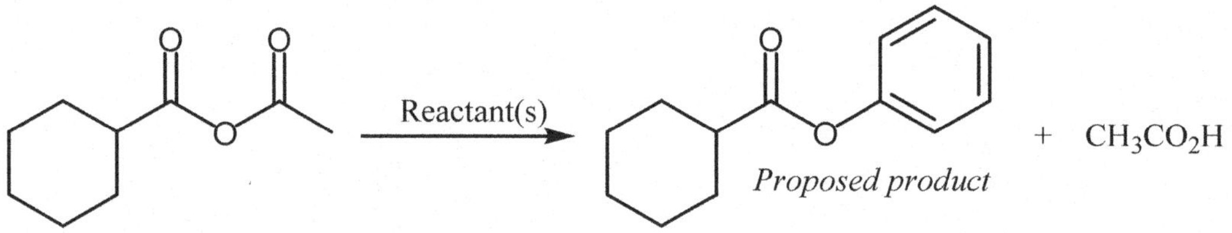

Solution 56.2

The above reaction is a nucleophilic acyl substitution where an anhydride is converted to an ester. If we circle the nucleophile in the reaction (as shown below) we can see it is a phenoxide (Ph—O⁻), so the missing reactant (NuH) should be PhOH with catalytic base.

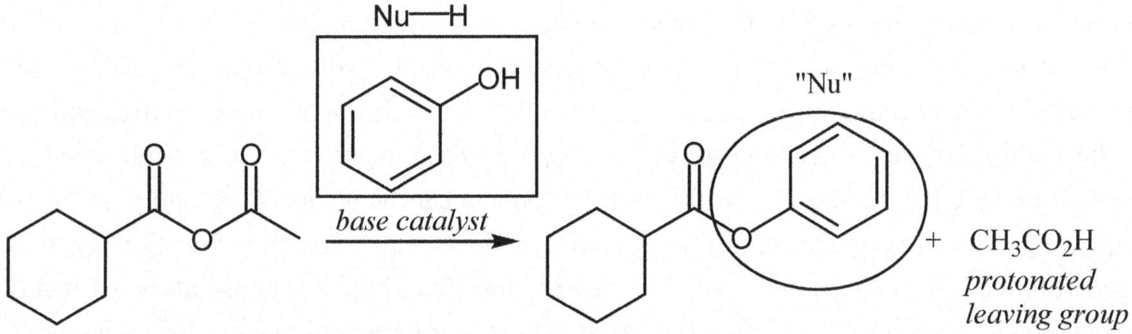

In this particular example, the two sides of the anhydride are different. One of the anhydride carbonyl carbons has a methyl substituent, whereas the other has a cyclohexyl substituent. Generally, we would predict that the nucleophile would attack the less hindered side. We would predict that the major product would result from nucleophilic attack on the acetate side and that the proposed product in the question would actually be the minor product:

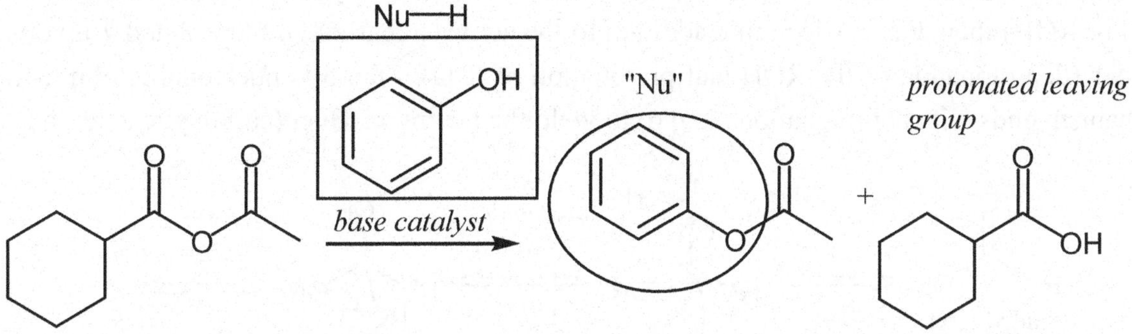

Lesson 57. Nucleophilic Acyl Substitution of Esters

Lesson 57.1. Hydrolysis/Transesterification

We learned in Lesson 55 that esters can be converted to carboxylic acids through nucleophilic acyl substitution, but esters cannot be converted to acid chlorides or anhydrides because this would require nucleophilic elimination of RO^-, which is less stable (a worse leaving group) than Cl^- or $RC(O)O^-$. We saw in Lesson 55 that esters can undergo acid-catalyzed hydrolysis (the reverse of Fischer Esterification). Esters can likewise be converted to other esters through **transesterification reactions**.

Like hydrolysis, the conversion of an ester into a different ester is a reversible reaction, which means that driving the equilibrium to favors the products can be achieved by applying LeChatelier's principle (e.g., adding more reactants, removing the products, etc.). Transesterification reactions follow two general mechanisms depending whether they are **base-catalyzed** or **acid-catalyzed** reactions. We are now going to study the acid-catalyzed transesterification of esters to make a different ether from one we already have.

<u>Acid-catalyzed transesterification/hydrolysis</u>:

This reaction mechanism is essentially the same as that of Fischer Esterification (Lesson 55). We also note that hydrolysis (reaction of ester with water to give carboxylic acid) is the reverse reaction of Fischer esterification. The first step is protonation of the carbonyl oxygen. Then, the neutral nucleophile (H_2O or ROH) then does nucleophilic addition to the activated carbonyl. Steps 3 and 4 involve proton transfer (deprotonation of the ROH and protonation of R'O). Step 5 is nucleophilic elimination. The mechanism ends with deprotonation (step 6) to yield the neutral product (carboxylic acid or ester):

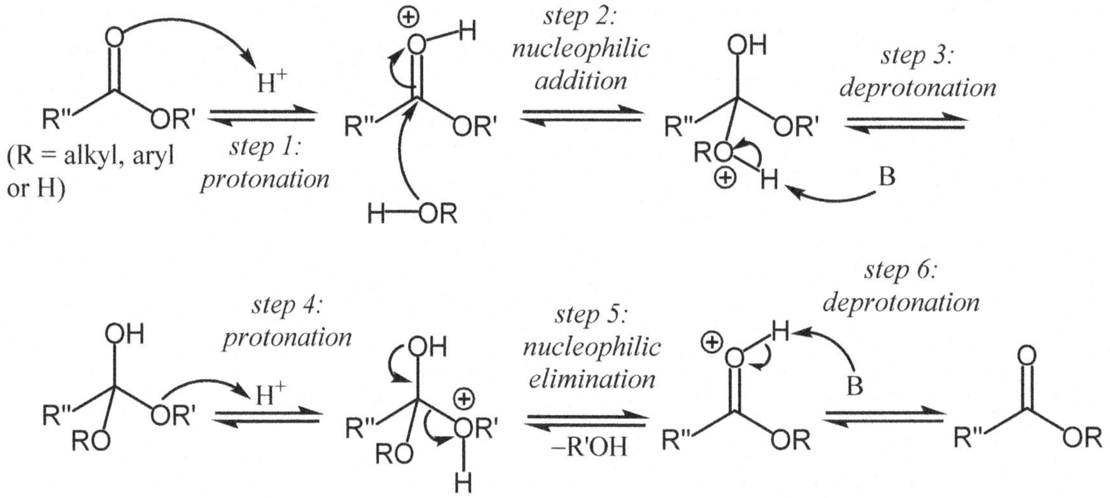

Lesson 57.2. Ammonolysis and Amidation Reactions of Esters

Esters can also be converted to amides upon reacting with 1° amines, 2° amines, or NH_3 under base- or acid-catalyzed conditions via mechanisms similar to the transesterification mechanisms:

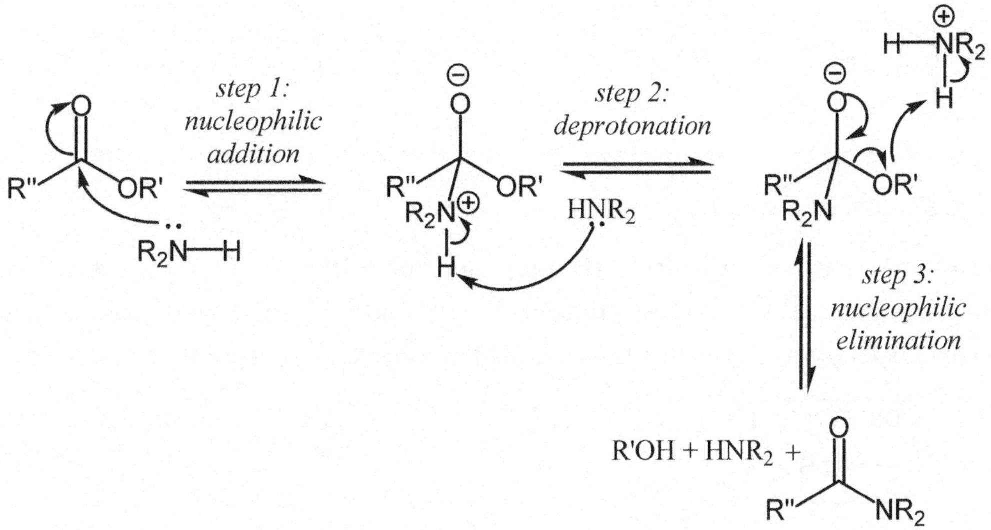

When ammonia is used as the nucleophile, this reaction is called **ammonolysis**. When a 1° or 2° amine is used, this reaction is called **amidation**.

Example 57.1

Give the missing reactant in this nucleophilic acyl substitution

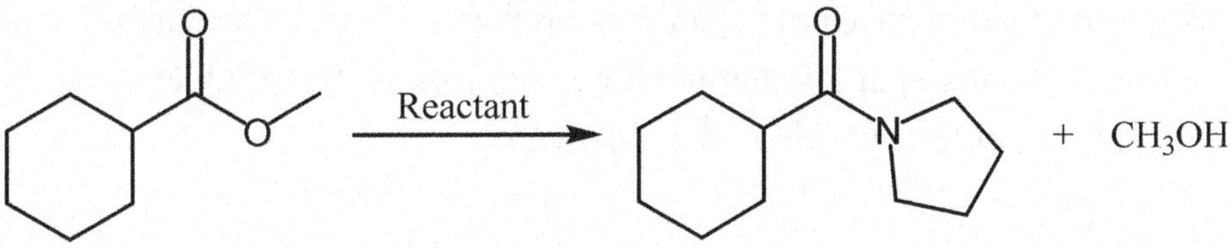

Solution 57.1

The above reaction is a nucleophilic acyl substitution where an ester is converted to an amide. If we circle the nucleophile in the reaction (as shown below) we can see it is a pyrrolidine anion so the missing reactant is NuH (amine)

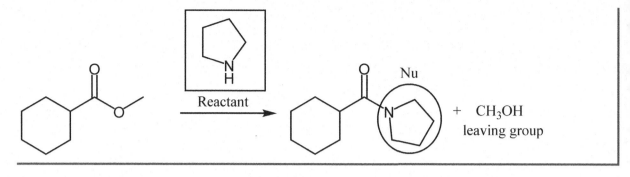

Lesson 57.3. Reduction of Esters to form Aldehydes

When an ester reacts with a hydride (H^-) as nucleophile the C—O π bond breaks forming the tetrahedral intermediate, then the oxygen reforms the π bond with the carbon and an alkoxy group is eliminated. This nucleophilic acyl substitution is also considered a reduction reaction.

The source of the hydride for this reduction reaction is Diisobutylaluminumhydride (DIBAL-H), a specialized metal hydride that we have not yet encountered in this book. The reaction mainly takes place in nonpolar solvents (like toluene) at low temperatures to promote the production of the aldehyde and prevent the further addition of the hydride to the aldehyde.

Lesson 58. Hydrolysis of Amides

Amides are the least reactive of the carboxylic acid derivatives, and they can only undergo hydrolysis reactions to give carboxylic acids under harsh reaction conditions. Hydrolysis of amides can take place either in very strongly acidic conditions (e.g. addition of aqueous HI) or very strong basic conditions (e.g. addition of concentrated aqueous NaOH). Furthermore, the reactions require heating to higher temperatures than are used for hydrolysis of esters.

The general mechanism in either acidic or basic media is similar to the corresponding mechanism for ester hydrolysis, with one exception: in an acidic medium, the amine (HNR_2) released by nucleophilic elimination (step 5) is a sufficiently strong base that it will deprotonate the cationic oxygen in step 6, to afford the ammonium $[H_2NR_2]^+$.

INDEX

Made in the USA
Monee, IL
05 February 2021